Bonobos
Encounters in Empathy

WRITTEN AND EDITED BY JO SANDIN

PUBLISHED BY THE ZOOLOGICAL SOCIETY OF MILWAUKEE
AND THE FOUNDATION FOR WILDLIFE CONSERVATION, INC.

I

Published by the Zoological Society of Milwaukee and
the Foundation for Wildlife Conservation, Inc.
10005 West Blue Mound Road, Milwaukee, Wisconsin 53226,
United States
www.zoosociety.org

ISBN - 13: 978-0-9794151-0-4
ISBN - 10: 0-9794151-0-1

First paperback edition, 2007

Designed by the Zoological Society Creative Department
Photography by Richard Brodzeller, Mark Scheuber,
Mike Nepper and the former Dierenpark Wassenaar Zoo

Cover: Murph, a male bonobo
at the Milwaukee County Zoo,
photographed by Richard Brodzeller

To all who pay attention and learn from the experience.

Bonobos: Encounters in Empathy

CONTENTS

Acknowledgments

It seemed like such a simple idea – giving the Zoological Society of Milwaukee (Wisconsin) a book telling the story of the Milwaukee County Zoo's stellar work with captive bonobos and the Zoological Society's equally outstanding bonobo conservation work in the Democratic Republic of Congo. Paula Brookmire, publications coordinator for the Zoological Society, was the first to see the idea's potential. She generously invested time and energy in launching the project. She believed the book could be, as I intended, an interesting Milwaukee story with global consequences. We also hoped that it could raise money both for bonobos at the Zoo and for bonobo preservation in Africa. Unfortunately, gifts of intellectual property such as manuscripts must be tightly wrapped in legal language to protect everybody concerned. Therefore, the book would never have been published without the help freely given by Milwaukee Attorney Paul Bargren in drawing up a contract that allowed the project to proceed. Both Paul and Paula have my sincere gratitude.

Nor is a gift of intellectual property easy to receive. Unlike a check, you can't just deposit it in the appropriate account. Once under way, the book depended on the work of scores of unselfish people who provided information, offered perspective, checked facts and/or proofread several drafts. After discharging their considerable duties at Zoo or Zoological Society, they uncomplainingly did their "homework for Jo." Any errors that remain in this book are all mine.

At the Zoo, this huge support team included Director Chuck Wikenhauser; Deputy Director Bruce Beehler; Jan Rafert, curator of primates and small mammals; Dr. Victoria Clyde, a Zoo veterinarian who also advises the Bonobo Species Survival Plan; Karin Schwartz, Zoo registrar; Patricia Khan, area supervisor for primates; zookeeper Mark Scheuber, whose photographs have so enlivened the prose; Claire Richard, gorilla zookeeper; and Jennifer Diliberti, public affairs director. Financial officer Vera Westphal set up a special bank account to receive the Zoo's half of the proceeds. Former zookeepers Rebecca (Becky) Loehe and Dr. Linda Cieslik were willing to share their memories and patiently proofread the resulting chapters. Leann Roth Beehler, who already had given the Zoo countless hours and expertise as she conducted prenatal ultrasound exams for pregnant bonobos, provided unstinting support for the endeavor.

At the Zoological Society, contributions of time and energy also began at the top with enthusiastic support from Chief Executive Officer Dr. Bert Davis and historical perspective from President Emeritus Dr. Gil Boese. They were willing to publish the book jointly through the Zoological Society and its partner, the Foundation for Wildlife Conservation, Inc., which Dr. Boese heads. Robin Higgins, vice president for communications, marketing and membership, employed her tact and charm as liaison for the project. Her assistant Lisa Bergmann and computer specialist Dominic Schanen distributed copies of successive drafts. Megan Ivers, Zoological Society spring 2006 intern, created a Web site activity on bonobos for teachers that can be found at www.zoosociety.org and compiled additional sources of information listed at the end of the book. Stephanie McLaughlin, conservation assistant, and Yvonne Walton, also of the conservation office, provided invaluable background material and issued updates on missions to the Democratic Republic of Congo. Richard Brodzeller, longtime freelance photographer, worked pictorial magic. Marcia Sinner, creative director, and her staff made the final result look gorgeous.

A thousand thanks go to these talented people along with my daughter and my husband, who helped me to keep this project in perspective, and to my colleague John Mollwitz, who was willing to cast a discerning eye over the final draft and suggest improvements.

My appreciation is unbounded for the bonobos themselves, the stars of this tale. I hope I have succeeded in capturing at least a part of their intelligence, their compassion and their engaging personalities. There would have been no story to tell had it not been for zookeeper Barbara Bell and Dr. Gay E. Reinartz, conservation coordinator of the Zoological Society, whose innovative work has improved the lives of bonobos in captivity and increased the chances of their survival in the wild. Both have devoted hours outside their already exhausting working hours to share their insights and experiences and to check the accuracy of this account. Dr. Harry Prosen, now retired from the Medical College of Wisconsin but still acting as consulting psychiatrist to the Zoo's primates, gave an extra dimension to the narrative. With a lifetime of work on empathy (and its absence) to inform him, he had both the ability and the authority to help us all to see what we were looking at – individuals of another species linked to each other and to us in empathy. I am so grateful.

Jo Sandin

The Bonobo Effect

Tamia, Laura and Ana Neema

The Bonobo Effect - an Introduction

Bonobos (bon-OH-bows) at the Milwaukee County Zoo –in the Midwestern U.S. city of Milwaukee, Wisconsin — have given new life to the idea that captive animals can act as envoys of their species, enlisting the sympathies and engaging the support of humans they encounter. To a degree that most human diplomats can only envy, these great apes have extended their influence around the world. They have

Photo provided by Hanneke Louwman and Sam LaMalfa

been a driving force in revolutionizing the way zoos in the United States and in Europe care for primates. (One even has his own psychiatrist.) They have been the impetus for an international effort to save bonobos in their central African homeland. (A former poachers' camp is now a research station.) The Milwaukee County Zoo has come a long way since 1986, when it welcomed seven frightened bonobos, suddenly transplanted to a new environment from the familiar nurture of a private facility in the Netherlands. By 2006 the Zoo was home to 21 bonobos, the zoo world's largest breeding group and one of the most socially authentic. In recognition of this achievement with "one of the most endangered and least studied of the great

Hanneke Louwman, one of the founders of the now-defunct Dierenpark Wassenaar Zoo in Holland, helped raise infant bonobos at the Zoo.

apes," the Zoo in 2006 was presented with the Association of Zoos and Aquariums' highest conservation honor, the Edward H. Bean Award.

Attempts to meet the needs of the bonobos in Milwaukee led to efforts to understand their condition and secure their preservation in the wild. The Zoological Society of Milwaukee – which has headquarters at the Milwaukee County Zoo – progressed from funding a few studies by graduate students to spearheading an internationally recognized conservation effort now under way in the Democratic Republic of Congo. A team put in place by the Zoological Society began the first systematic survey of the wild population in Salonga (sah-LOHN-gah) National Park, a UNESCO World Heritage

Photo provided by Gay E. Reinartz

Deep in the African forest of Salonga National Park Dr. Gay Edwards Reinartz and the Zoological Society's Congolese staff search for bonobos. From left are Dr. Reinartz, Mboyo Bolinga, Nduzo Bokono-Bolungi, Botomfie Mompansuon, Isomana Edmond.

Site and one of only two areas specifically set aside for bonobo preservation. The Zoological Society plays an active role in turning protective laws into actual defense against poachers. Congolese who staff the research station at Etate (ay-TAH-tay) at the northern tip of the Salonga report the surest sign that their anti-poaching partnership with the Zoological Society is beginning to succeed: frequent sightings of wild bonobos feeding, playing and rearing their young in the equatorial rain forest that is their natural home.

Conditions of daily life among captive bonobos have improved and chances for survival among their wild relatives have increased because two species of primates – humans and bonobos – paid attention to each other and learned from that experience. Since bonobo social structure is matriarchal in character, it is fitting that this adventure in discovery was launched by three females: Maringa, for two decades leader of bonobos living at the Milwaukee County Zoo; Barbara Bell, head keeper for bonobos; and Dr. Gay E. Reinartz, conservation coordinator for the Zoological Society of Milwaukee and coordinator of the Bonobo Species Survival Plan for the Association of Zoos and Aquariums. This book tells a part of their ongoing story.

Photo by Richard Brodzeller

Maringa the bonobo matriarch is bald from repeated but respectful grooming by the other bonobos.

Below:
Zookeeper Barbara Bell (left) rewards a bonobo (lying above her for easy and safe access) for letting her abdomen be rubbed for an ultrasound.

Photo provided by Gay E. Reinartz

Dr. Gay E. Reinartz trains Mboyo Bolinga in using Global Positioning System (GPS) technology.

Photo by Richard Brodzeller

Bonobos: Encounters in Empathy

Rarest of the Rare

Zanga Mokila

CHAPTER 1.

Rarest of the Rare

Bonobos – with their flamboyant personalities and intricate sexual politics – seem to be perfect candidates for celebrity roles in the primate world. The daily lives of bonobos (*Pan paniscus*) play out like soap operas. So why is the bonobo the species of great ape most likely to be ignored by humans? In his 1997 book, Frans B.M. de Waal, professor of primate behavior at Emory University in Atlanta, Ga., and one of the foremost students of the species, calls the bonobo "The Forgotten Ape."

Photo by Richard Brodzeller

There's a simple explanation. Even in the world of endangered species, bonobos are rare. Their fellow great apes outnumber them in international zoos. In 2006, there were a mere 84 bonobos at 10 zoos and research centers in North America. Only Columbus, Cincinnati, the Great Ape Trust of Iowa, Fort Worth, Jacksonville, Memphis, Milwaukee, San Diego Zoo and San Diego Wild Animal Park, and Morelia (Mexico) host these uncommon primates. Nine European zoos were home to even fewer bonobos, only 76, in 2006. At that time another 47, most

Dr. Frans B.M. de Waal (left) views Milwaukee's bonobos with Jan Rafert, the Milwaukee County Zoo's primate curator.

of them orphans rescued from the pet trade, had found refuge at Lola ya Bonobo (Paradise for Bonobos), a private sanctuary near Kinshasa, capital of the Democratic Republic of Congo (formerly the Belgian Congo, then Zaire). The vast equatorial rain forest south of the Congo River in this central African nation is the one place on Earth that bonobos live outside captivity. No one knows how many of this endangered species remain there. The Zoological Society of Milwaukee, in cooperation with the Congolese Ministry of the Environment and the Institut Congolais pour le Conservation de la Nature (ICCN), is still engaged in the first systematic attempt to assess the population of wild bonobos in Salonga National Park, one of two federally protected areas for the species.

Also, it's hard for an endangered species to make a name for itself when other great apes are media stars. Consider this:

"Gorillas in the Mist," the story of researcher Dian Fossey, is a major motion picture with Sigourney Weaver.

"Jane Goodall's Wild Chimpanzees" is a perennial television favorite.

Although Birute Galdikas is not as widely known for her work in orangutan conservation in Borneo and Sumatra, the Asian apes themselves appear on camera in company with actress Julia Roberts.

IT'S NOT EASY BEING SEEN

By contrast, bonobos are so little recognized that references to them inevitably begin with comparisons to their better known cousins, the common chimpanzees.

Indeed, the formal designation for the species remains "pygmy chimpanzee," a reference to the apes' original assessment as being somewhat smaller than chimpanzees (*Pan troglodytes*). Standard computer spell-check programs still do not recognize the word "bonobo." There is no agreement on the origin of that name for the species. Some suggest that it is a misspelling of Bolobo, a town on the Congo River. Others say it comes from the ancient Bantu language word for "ancestor."

Although Tulp's ape, the first imported to Europe in 1641, probably was a bonobo, the species was not recognized as such for another three centuries. Then in 1929, a researcher looking over a collection of skulls at a Belgium museum proposed that one of them belonged, not to a juvenile chimpanzee, as had been previously thought, but to an adult. This led him to conclude that the specimen represented a separate subspecies of chimpanzees. A more extensive examination of a greater number of skeletal specimens moved Harold Jefferson Coolidge, assistant curator of mammals at Harvard's Museum of Comparative Zoology, to go one step further. In 1933, he successfully proposed that the skeletons be reclassified as a full species, *Pan paniscus*. Humans finally had recognized a divergence which, according to current thought, began a few million years ago when chimpanzees moved out of the forest and into dry open areas, and bonobos mainly stayed in the trees.

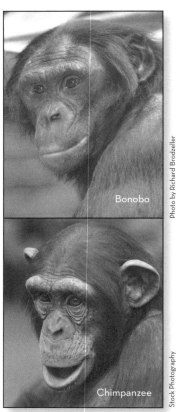

Bonobo

Chimpanzee

Photo by Richard Brodzeller

Stock Photography

Superficially, there remain many similarities between the two species. Both primates qualify as humanity's closest living relatives, sharing 98.4% of the DNA that characterizes *Homo sapiens*. The split between the ancestor of modern humans and the common ancestor of bonobos and chimpanzees may have occurred no more than eight million years ago. There are a number of clear differences between the two apes. Bonobo bodies are more slender, with longer, thinner arms, sturdier legs and narrower shoulders. Their skulls are smaller, more rounded than those of chimpanzees. Brow ridges are smaller, jaws less protruding. They are more likely to walk upright. Bonobo hair is black, rather than brown, with distinctive tufts over their ears and a neat part in the middle. Hands and feet are dark colored, in contrast to the lighter coloration of chimpanzees.

However, the greater contrast is social. Unlike the male-dominated groups that characterize chimpanzee social order, bonobo groups tend to be more matriarchal in structure. Status generally is determined by relationship to one of the dominant females. "Make love, not war" seems to be the guiding ethic. Sexual contact is used to create and cement bonds within the group, to relieve tension and to make peace with offenders of community rules. Affectionate contact including hand-holding, kissing, genital rubbing and frontal copulation has been observed not only between males and females but between females and sometimes between males. It is the females, roughly at an age when they reach sexual maturity – around 8 years old in captivity, 10-12 years in the wild – who leave the family group to form alliances elsewhere. Sons generally remain with their mothers' group where their status is related to that of their female relatives. No careful observer paying attention to a reasonably normal group of captive bonobos would confuse them with chimpanzees.

Seeing and Seizing an Opportunity

In 1986, Jan and Hanneke Louwman, owners of the Dierenpark Wassenaar, a private zoo near Amsterdam, sought a new home for all seven of their bonobos. To ensure that any move would involve the least possible stress for the primates, they wanted a single placement for the entire group. Dr. Gil Boese, then director of the Milwaukee County Zoo, and Dr. Gay E. Reinartz, then Zoo registrar, saw an enormous opportunity not only to place bonobos in Milwaukee for the first time, but also to increase the breeding vitality of the entire bonobo population in the United States. At that time, the only bonobos in the country were 12 at the San Diego Zoo in California and 10 in Georgia – at Yerkes National Primate Research Center and the language laboratory at nearby Georgia State University. Together both locations had only four wild-caught animals. All the rest had been bred in captivity. For the gene pool of a proposed captive-breeding program that would encompass North America, the Dierenpark bonobos, including five captured in what was then called Zaire, represented a potential reservoir of essential diversity. In retrospect, there is only one possible name for the group as a whole:

The Magnificent Seven

By December 3, 1986, Milwaukee was the new home of a group destined to become stars of the zoo world. The bonobos (with their ages in 1986) were:

Kidogo (kih-DOH-goh), a 12-year-old male who died in 1996 of heart disease and whose end-of-life care by his friend Lody became a powerful statement of primate empathy;

Lannie, a 32-year-old female, who died in 1987;

Kitty, a 36-year-old female who now, at 57, is one of the oldest bonobos living in captivity;

Lody (LOH-dee), a 13-year-old male whose strong leadership as dominant male has helped ensure the social health of the Milwaukee group;

Maringa (mah-RING-gah), a 14-year-old female whose mothering style has been a model for the entire group and who has been dominant female in this matriarchal society for most of her life here;

Lomako (loh-MAH-koh), the 2-year-old son of Lody and Maringa, whose quick responses to Bonobo School have enhanced the success of that endeavor;

Naomi, the 4-year-old daughter of Lody and Maringa, believed to be the first bonobo delivered by Caesarean section. She died in 1990.

With their arrival began a series of events that have influenced not only the way people care for other species in captivity and in the wild but also the way we human beings understand ourselves.

Divas and Such

Baby Diedre and Kosana

Chapter 2.

Divas and Such

Powerful personalities – only some of them human – have shaped the great-ape world in the last 20 years. During those two decades the Milwaukee County Zoo became home to the largest breeding group of bonobos outside Africa, and the Zoological Society of Milwaukee teamed with the Congolese government to begin groundbreaking conservation work in the bonobos' homeland. Human players tell of their parts in these dramatic developments in their own words. Unfortunately, bonobo language is not yet accessible to us. However, the primates *have* communicated eloquently in behaviors and emotions that are immediately recognizable: tender affection, empathetic concern, towering rage, manipulative tantrums, humble apologies, nurturing love. Therefore, it is important to our understanding of this story to meet those individuals whose lives and characters are central to everything that has happened: the bonobos themselves.

Founder animals – those captured in the wild and bred in captivity – bring vital genetic diversity to the entire North American captive bonobo population. However, most of Milwaukee's founder animals also contribute equally important gifts, ingrained patterns of natural bonobo behavior that have allowed Milwaukee's bonobos to continue to act, in captivity, much like their cousins in the rain forest.

The Royal Couple

Photo by Mark Scheuber

Maringa and infant Faith

Maringa has been matriarch, the ultimate power in the group, since she arrived Dec. 3, 1986, and is only now beginning to relinquish her position. Two physical characteristics make her immediately recognizable. Because she has been the principal subject of grooming (an expression of admiration, affection and even servility) for so many years, she has less body hair than any other bonobo. She has a pronounced limp, although neurological tests have offered no reason for this condition.

She was born in what was then Zaire (now the Democratic Republic of Congo) sometime in 1972. (All birth years for bonobos captured in the wild are estimates.) Captured in her youth, she came to Milwaukee with her longtime mate, Lody, via the former private zoo Dierenpark Wassenaar in the Netherlands. Dutch sailors had purchased the youngsters and sold them again after docking in Holland, a practice that was not uncommon at the time. Zoo owner Jan Louwman and his wife, Hanneke, who saw their private zoo as a kind of rescue operation, showered the infant newcomers with love and attention. In retrospect their nurture is credited with mimicking the kind of doting care provided by bonobo mothers and allowing the youngsters to develop, insofar as is possible in captivity, as normal bonobos. Obviously, the bonobos themselves remember that time. When Hanneke Louwman visited Milwaukee almost 20 years after the bonobos had left the Netherlands, Maringa recognized the Dutch woman as her "foster mother." Maringa responded with noisy demonstrations of delight when Louwman sang a song she had used to calm the infant bonobos.

Maringa is mother to Lomako, Zomi and – most recently – Faith. A previous daughter, Naomi (who died in 1990), was born at Wassenaar in 1982 in what is believed to have been the world's first Caesarean section delivery of a bonobo. Another daughter, Eliya, was born at the Milwaukee County Zoo December 17, 1990, and died of pneumonia at age 7.

Lody

Lody until recently was the group's alpha male. Wild-born in 1973 in Zaire (DRC), he was captured as a youngster and brought to the Netherlands by Dutch sailors, who sold him and Maringa to the Dierenpark Zoo. For two decades he has served as prince consort in the influence-wielding style of England's Prince Albert. He was purchased by the Milwaukee County Zoo in 1986 and arrived in Milwaukee December 3 with Maringa and their son, Lomako. Also with the group were Kidogo, a wild-caught male about Lody's age who died in 1995; Kitty, an elderly female; and Lannie, another wild-caught female from Wassenaar who later died in childbirth. Psychiatrist Harry Prosen, a consultant to the Milwaukee County Zoo for bonobos and other animals, and keeper Barbara Bell credit Lody's wise leadership with the high level of emotional health of Milwaukee's bonobos and points to numerous instances of empathetic behavior toward other members of the group. His days of mourning after the death of Kidogo constitute one of the most poignant chapters of the community's life. He is father to Lomako, Zanga Mokila, Zomi, Claudine and Deidre.

FOUNDING MOTHERS (AND FATHER)

Founder animals are those captured in the wild and successfully bred in captivity. Wild-caught animals not producing live offspring and, therefore, contributing to the gene pool are not considered to be founders.

Linda, captured in Zaire (DRC) as an infant, has been a power in the group ever since she arrived in Milwaukee on loan from the San Diego Zoo. Along the way, she lived for 13 years at Yerkes National Primate Research Center

Photo by Richard Brodzeller

Linda

14

at Emory University, Atlanta, Ga., which no longer has bonobos. She came to Milwaukee in 1995. Since Linda's birth year is estimated at 1954, she is second only to Kitty in age among the Zoo's bonobos. Although she has medical challenges (diabetes), Linda has been second only to Maringa in dominance among the group's powerful females. When rebuking the much larger and younger males, Linda characteristically bloodies the offenders and emerges unscathed from the encounter. After a lifetime of bearing infants – three sons, nine daughters – removed from her almost at birth, Linda has proved to be a brilliant caretaker. She has provided foster care to young bonobos raised elsewhere apart from their mothers and sent to Milwaukee for socialization. Among her successes is her nurture of her great- grandson, Zuri.

Photo by Richard Brodzeller

Kosana

Kosana (koh-SAH-nah), the youngest of Milwaukee's founder animals, arrived here after an odyssey worthy of tabloid headlines. Her estimated birth year is 1982. She was captured in Zaire (DRC) in 1983, evidently before she was able to learn many of the complex social graces that character-ize bonobo life. Until she found her place in Milwaukee, she was described as socially clueless. Imported into Belgium before that nation was a party to the Convention on International Trade of Endangered Species on Wild Fauna and Flora (CITES), she was sold to the Japan Monkey Centre and Primate Zoo June 1, 1990. News of the sale pro-voked such a huge public protest from Japanese conservationists that she was returned to Belgium to be placed in the Antwerp Zoo in 1992. By July 18, 1994, she was transferred on recommendation of the European species survival plan for bonobos to the Leipzig Zoo, where she gave birth twice but raised neither of her offspring. Returned to Antwerp from Leipzig July 21, 1999, she was donated to the Planckendael Animal Park, run by the Royal Zoological Society of Antwerp.

On August 1, 2000, Dr. Gay E. Reinartz, coordinator for the Bonobo SSP in North America, recommended that Kosana come to Milwaukee in hopes that bonobos here could teach her social and parenting skills. Her patient keeper, Joris Jacobs, spent a year preparing her for her new home, working carefully with her in English, so the new language would not be a shock. To minimize the trauma of the long journey and trip, the plan was for the pregnant Kosana to travel from Leipzig with another female bonobo, Unga, bound for the Columbus Zoo, where they would spend their required quarantine together.

Then the tale became a cliffhanger. Their flight left Europe September 11, 2001. The terrorist attack on the World Trade Center in New York City and the Pentagon grounded all air traffic in the United States. The plane was diverted to Gander, Newfoundland, where all passengers – including bonobos – were kept on board for 13 hours. The two bonobos were transferred from the cargo hold and housed in their crates at a small hanger at the airport until flights resumed September 16. On that day, the bonobos were flown to Chicago and then trucked to Columbus, arriv-ing at 3 a.m. September 17. During her quarantine there, Kosana spontaneously aborted the fetus on October 11.

She finally arrived in Milwaukee October 24. Six days later, a Milwaukee keeper observed the crouched-over walk that Antwerp keepers had described. "Kosana does this when nervous," the Milwaukee keeper noted. "She acts as if the sky is falling." Small wonder. However, since arriving in Milwaukee, Kosana has found a place in the group. After giving birth March 4, 2003, she is successfully raising a daughter, Deidre, sired by Lody.

Photo by Mark Scheuber

Viaje (vee-AH-hay), whose name means "journey" in Spanish, is the most recently arrived of Milwaukee's founder animals. He was born in Zaire (DRC) in 1980, and is on loan from Zoofari, a private zoo in Cuernavaca, Mexico, about 50 miles south of Mexico City. Wild-caught as an infant of about 1 year, he was imported to Mexico from Belgium in 1981 at a time when neither country was party to CITES. In Cuernavaca, he lived with a female named Chispita (chis-PEET-ah), also imported as an infant, until 1993, when his mate died. Immediately, his

Viaje

owner, Marcos Orteiza, began seeking a new home for him. Since bonobos are such social animals, Orteiza knew that remaining alone would be damaging for the young male. Milwaukee was willing to take him, but getting him here took five years, mountains of paperwork in English and Spanish, and a journey of 1,751 miles. He finally arrived at the Milwaukee County Zoo November 28, 2001, after last-minute glitches, amazing interventions and quarantine at the St. Louis Zoo. Although his adjustment to the group was tumultuous, he already has fathered one infant, Maringa's latest daughter, Faith.

THE BOLD AND THE BEAUTIFUL – POWER FEMALES

Kitty, at 57, is probably the world's oldest captive bonobo. Although she was born in the wild, she has never given birth, so she is not considered to be a founder animal. Since her birth year is estimated to be 1950, she first took breath in

Photo by Mark Scheuber

equatorial Africa 10 years before the Congo became independent of Belgian colonial rule. (Another female, Margrit, at the Frankfurt Zoo, has an estimated birth year of 1951 and could have been born before that year. However, Margrit did not appear in zoo records until 1959, five years after Kitty.) Kitty arrived in Milwaukee in 1986 at the same time as Lody, Maringa and Lomako. She is on loan from Planckendael Zoo near Antwerp, Belgium. On the recommendation of the Bonobo Species Survival Plan, Kitty was transferred from Milwaukee to Yerkes in 1988 for a stay of several years. She returned to Milwaukee in 1995. Now blind and deaf as well as subject to seizures, she nonetheless remains of enormous importance to the group both as a giver and a

Kitty

recipient of compassionate care.

Bonobos: Encounters in Empathy

Laura

Laura, born August 27, 1967, at the San Diego Zoo, is still on loan from that institution. Since her arrival in Milwaukee December 16, 1993, she has joined her mother, Linda, as one of the group's dominant females and an important member of the captive-breeding community as well as a talented foster mother. She came to Milwaukee from Yerkes National Primate Research Center at Emory University, Atlanta, Ga., along with her son, Murph, who was born at Yerkes. Here in Milwaukee she has given birth (July 25, 1995) to Yatole, (yah-TOW-lee), a female with chromosome damage who died in 10 months, and two healthy daughters: Zanga Mokila and Claudine. Laura has served as foster mother to Makanza, whose mother, Zalia (zah-LEE-ah), died in 1995 when he was only 16 months old, and to Zuri, her great-grandnephew.

Ana Neema (AH-nah NEE-mah), was captive-born February 15, 1992, at the Language Research Center at Georgia State University where she lived until coming to Milwaukee in July 1999. A lightning-fast learner, she showed immediate interest in training. Before she had been in Milwaukee a week, she volunteered to participate in a click-and-reward training session for her breakfast, earning favorite foods by responding correctly to verbal commands reinforced by the sound of a clicker. In the first three weeks, she learned "right" and "left" and mastered colors. Reared by her mother, Matata, for her first two years, she has all the

Photo by Richard Brodzeller
Ana Neema

nuances of bonobo etiquette down pat and has become a powerful enforcer of group behavior standards. At Milwaukee she has given birth to a son, Bila Isia, in 2001 and a daughter, Gilda, February 6, 2006.

Photo by Richard Brodzeller
Tamia

Tamia (tah-MEE-ah), Milwaukee's youngest power female, was born July 5, 1996, at the Columbus Zoo. She learned bonobo social graces from her mother and other members of the Columbus bonobo group, which numbers 16, the second largest in the United States. Since she arrived in Milwaukee December 15, 2004, she has ingratiated herself with the ruling sisterhood and already has become a femme formidable in the tradition of the conniving ingénue who replaces the established star.

The Young and the Restless –
Contenders for the Consort's Crown

Although Milwaukee bonobo society, like that observed among wild bonobos in DRC, is matriarchal in its power structure, the alpha male fulfills an important role in the life of the group. Females definitely (and sometimes violently) decide for themselves with whom they will mate. However, high-status males often are given preference. Also, the alpha male helps set the tone of community life and curb the obnoxious tendencies of adolescent males. Since the mother-son bond is the most important one in group continuity and since sons usually remain in their mothers' groups for life, a male supporter is useful to enforce community standards.

Lomako

Lomako was born to Lody and Maringa while they were at Wassenaar on June 25, 1984. As Maringa's son in a group in which a mother's status determines the rank of offspring, he is one of the contenders for his father's former position but by no means a shoo-in for the role of alpha male. He arrived in Milwaukee in 1986 with his parents. When bonobo-training sessions began in the early 1990s, Lomako distinguished himself as an exceptionally quick learner. He easily mastered behaviors designed to allow keepers to monitor blood pressure, draw blood samples and administer cardiovascular scans. He taught himself to jump on cue and performs a dandy back flip. Along with Ana Neema, he was among the first to recognize symbols for desired items. He enjoys painting, and one of his works was shown on the "Late Night With David Letterman" TV show in August 2005. Like his father, Lody, Lomako has exhibited gentle, compassionate behavior, especially toward the aged Kitty. On the other hand, he frequently has teased Linda past the limits of her patience. Although he often has acted as deputy to Brian in the ongoing skirmishes to be recognized as No. 1 Male, Lomako was severely wounded in a fight with Brian in early 2005. However, the battle for the position of the group's male (and therefore secondary) leader is by no means settled. As his father's status has diminished due to Lody's failing health, Lomako has provided quiet companionship to Lody.

Brian, who arrived in Milwaukee from a research facility in 1997, was born January 8, 1989. As the first bonobo known to

Photo by Mike Nepper

Brian

have his own psychiatrist, he has made international headlines with articles in the Sunday Times of London and *Der Spiegel*, to name only two. In 1997 Dr. Harry Prosen, who was then head of the psychiatry department at the Medical College of Wisconsin, began to treat Brian for extreme anxiety, which caused the young male to harm himself and to induce vomiting. There never has been any doubt about Brian's intelligence. Within months of arriving here, he

Dr. Harry Prosen

had learned the words "hands," "back" and "feet" on his own. By watching other bonobos at training, he learned more behaviors. A week after he was introduced to colors, he reliably recognized red, yellow, blue and white. However, he suffered from a gigantic knowledge deficit about bonobo relationships. Clueless about the culture's complex etiquette, he repeatedly provoked the dominant females. They bit him bloody on more than one occasion and still administer beatings when they consider that he has overreached himself. Gradually, however, he has learned enough about group standards to reconcile successfully with even a power as great as Linda. He also has risen in status. In March 2002, he was ranked third behind Lody and Lomako. By April 2002, he had strutted into position as No. 2 Male. In 2005, he and Lody reversed roles as the longtime male leader began to withdraw from group politics.

Murph came to Milwaukee December 16, 1993, with his mother, Laura, as an infant. He was born April 15, 1990. Developing right on schedule for young bonobos, he entered the brat stage at age 9, pestering and teasing any bonobo within sight or reach. No matter what the grouping, he made himself obnoxious for several years and still does so on occasion. Then in March 2003, he attracted the interest of matriarch Maringa after years of being on the outs with her. By the following January, he had parlayed that relationship, his

Photo by Richard Brodzeller

Murph

mother's influence and his increased size into a bid for alpha position, jockeying for status with Brian. In July 2005, however, he received a humbling lesson on just who actually wields power in the group. He tried to take on the females without Laura's backing and they ripped his fingers and toes. It ain't over yet.

SIDEKICKS

Makanza

Photo by Mark Scheuber

Makanza (mah-KAHN-zah), born August 11, 1994, accompanied his mother, Zalia, to Milwaukee in November 1995, but she died under anesthesia during a quarantine health workup in December of that year. Fostered by Laura for five years, he "traded mothers" in 2000, shortly after Laura's son Murph entered the annoying stage of adolescence. Makanza sought and received nurture from Maringa, who welcomed him into her care despite the fact that she still had infant Zomi as her primary focus of concern. He is too young to contend seriously for alpha male status. However, he already has begun siring progeny (Bila Isia in 2001 and Gilda in 2006). He also mixes it up with the major players and gains influence by lending support to those competing for dominance.

Zuri (ZUR-ee) arrived in Milwaukee April 5, 2000, for socialization. Rejected by his mother at birth June 10, 1998, he was hand-reared by keepers at the San Diego Zoo, where he had never been held or carried by another bonobo. As soon as he was released from quarantine here, Zuri, 22 months old, got a warm welcome from the bonobos. Lody not only carried, cuddled and hovered over the newcomer, but the alpha male also made certain that the

Photo by Richard Brodzeller

Zuri

"security blanket" to which the infant clung went everywhere that Zuri went. Like Makanza, he can serve as friend to a contender as older males rearrange the group's pecking order.

ALL OUR CHILDREN – MILWAUKEE COUNTY ZOO BABIES

Photo by Richard Brodzeller

Zanga Mokila

Zanga Mokila (ZAHNG-ah mo-KEE-lah) was born January 9, 1999, at the Milwaukee County Zoo to Laura. Zanga's name, in Lingala, a Congolese trade language, means "without a tail." Her father was Lody. Only hours after Zanga's birth, Laura willingly allowed keeper Barbara Bell to touch the infant, an amazing demonstration of trust. As a toddler, Zanga showed how important bonobo babies are to the social and emotional well-being of the group. Under her mother's watchful eye, Zanga enjoyed infant play – laughing, tickling and cuddling – with Lody, Brian and later with Viaje. Her presence was vital to the welcome her mother provided to a foster child, Zuri.

Zomi (ZOH-mee) was born July 17, 1999, at the Milwaukee County Zoo to Lody and Maringa, the quintessential possessive Supermom. When Zomi first began venturing out on her own January 19, 2004, Maringa flung herself on the floor whimpering and protesting. Zomi herself began separating from her mother without a lot of fuss. Soon she was at home with other youngsters in a group supervised by her father. She also adjusted fairly quickly to having to fight her own battles on the playground.

Photo by Richard Brodzeller

Zomi

Photo by Richard Brodzeller

Bila Isia

Bila Isia (BEEL-ah EESS-ee-ah) was born August 14, 2001, at the Milwaukee County Zoo and named after Inogwabini Bila Isia, then field conservation director for the Zoological Society's bonobo conservation project in Congo. His mother is Ana Neema and his father, Makanza. By July 6, 2003, before Bila Isia was 2, he began to join in the training sessions when keepers worked with his mother.

Claudine was born at the Milwaukee County Zoo August 23, 2002, to Lody and Laura. By March 28, 2003, Claudine already was beginning to scoot away from her mother. By September 28, 2003, she was starting to play rambunctiously with the other infants. Claudine was named after Claudine Andre, who runs a bonobo sanctuary near Kinshasa, capital of DRC.

Photo by Richard Brodzeller

Claudine

Photo by Mark Scheuber

Deidre

Deidre, born at the Zoo March 4, 2003, may have been two or three weeks premature. Her mother, Kosana (who came from Belgium), had been described as socially and maternally inept by European zookeepers. Still, Milwaukee's staff elected to keep the infant with Kosana rather than intervening to remove the baby. The "gamble" paid off. In three weeks, Deidre was the size of a normal newborn. Acting on advice from Milwaukee (Wis.) lactation

Bonobos: Encounters in Empathy

consultant Pat Gima, keepers gave Kosana an enriched diet that had the expected result of improving Kosana's milk supply and the baby's diet. By September 10, 2003, the infant was described as "very bright and smiley." The Zoo had another experienced mother in its group. DNA testing revealed that the infant was sired by Lody.

Faith, Maringa's most recent daughter, sired by Viaje, was born February 19, 2005. Although she and her mother already have gotten caught in the power struggle that is currently identifying the group's new leaders, she so far has weathered the storm. Her father, Viaje, is too low-ranking to compete for the position of alpha male or to offer much protection. Genetic diversity, which he brings to the captive-breeding program, is something that matters more to SSP coordinators than to bonobo movers

Faith

and shakers. Therefore, one of the contenders, Brian, frequently roughs up Faith and Maringa in hopes of provoking his genuine rivals, Lomako and Lody.

Gilda

Gilda, born to Ana Neema and Makanza February 6, 2006, is still an unknown quantity.

Mixing Volatile Chemicals

Maringa being groomed by Zomi

Chapter 3.

Mixing Volatile Chemicals

While most other primates live in stable communities in which membership remains constant, bonobos, chimpanzees and spider monkeys like to regroup frequently. Even their long-term communities, reflecting the life-long bond between mother and son, are constantly in flux. Scientists call these fission-fusion societies. At the Milwaukee County Zoo, keeper Barbara Bell considers the description particularly appropriate because it connotes explosive energy. She likens her daily efforts to combine individuals in compatible groups to "mixing volatile chemicals." Managing the zoo world's largest community of captive bonobos is a little like being a diplomat in a country where you don't speak the language. People have just begun to understand these fellow primates in the 70-some years since bonobos were recognized as a separate species. Although bonobos have shown amazing aptitude for recognizing human language, humans have yet to decode any but the most obvious communications of these great apes.

We know so little about bonobos in their natural home because it is difficult, expensive and politically challenging to conduct systematic surveys in the vast and remote rain forest where they live. The Zoological Society of Milwaukee is conducting assessments in the Salonga National Park of the Democratic Republic of Congo. Researchers caution, however, that any current estimates of how many bonobos might live there must be considered as preliminary "best guesses." Yet somehow bonobos have survived a series of cataclysms that continue to push them toward extinction: two consecutive civil wars in which as many as 4 million humans may have died; the economic collapse of their homeland; capture for the pet market; extensive poaching for the bushmeat trade; and habitat destruction by loggers and slash-and-burn farmers.

Photo provided by Gay E. Reinartz

Zoological Society staff members (from left) Mboyo Bolinga, Nduzo Bokono-Bolungi and Guy Tshimanga collect GPS data.

In 1973 Takayoshi Kano and other researchers from Kyoto University established a research station at Wamba in the northern part of the bonobo homeland outside Salonga National Park. Before civil unrest drove the Japanese away in 1991, they had identified as many as 200 individuals lured from the rain forest by sugar cane. After the civil wars, researchers returned to Wamba in 2002 on a regular basis. They found that though there had been losses to poachers, most members of the main study group still were alive. Nevertheless, one of the most alarming developments observed by Wamba researchers over the entire course of their work was the severe erosion of a taboo among villagers against killing bonobos.

Although we may not know their numbers in natural communities, we do know that bonobos risk extinction and that their continuance as a species may depend in part on the health of the captive community. The Association of Zoos and Aquariums' Species Survival Plan (SSP) for bonobos, developed in 1988 and coordinated by Dr. Gay E. Reinartz, conservation coordinator for the Zoological Society of Milwaukee, depends for success on the continued contributions of as many genetically diverse individuals as possible. The SSP selects breeding partners based on their genetic relationship to each other and to the population as a whole. This necessity plays havoc with the norms of natural bonobo society. Japanese researchers have observed that in the wild, male bonobos stay with their mother's group for life while females migrate to other groups during adolescence. In a captive-breeding population, it isn't always possible to follow this model, however sensitive are the custodians of the Species Survival Plan. To mimic natural social groupings, SSP recommendations usually relocate young adult females with a group outside the one into which they were born, but the situation with young males is more difficult.

In the captive gene pool, an individual's value is determined by the genetic diversity he offers, not by the rank of his mother. Therefore, a number of unrelated males find themselves introduced to each other (and to the dominant females of a particular group) as adults or adolescents, a situation that leads to inevitable tensions. Even individuals whose life history has left them deficient in bonobo social graces must be integrated into a successful breeding program. That makes it easier to understand why Barbara Bell talks about "volatile chemicals" when she mixes and matches the bonobos in her care. Bell has been so successful in integrating so-called problem bonobos into Milwaukee's community that she has further complicated her own work. Because genetically valuable, but socially inept bonobos so often have found a place in Milwaukee, the Zoo now shelters an unusually high number of those individuals. All this is good for species survival, but it makes zookeeping a daily adventure.

In the rain forest, a bonobo that offends group rules – for example, by snatching a choice food item from a higher ranking companion – can simply be driven into temporary exile in a neighboring tree. The same fate curbs the obnoxious behavior of young males in what can only be called the "brat stage" of their development, beginning about age 9 and often continuing for a decade, sometimes longer. The lifetime bond between mother and son can be maintained at the same time that peace is preserved in the group as a whole. Captive bonobos have limited space in which to express their preferences and work out their differences. Therefore, Bell and other keepers must pay close attention to combining individuals – on exhibit, in the off-exhibit outdoor enclosure, in the playroom and in sleeping areas – in ways that reflect current group dynamics.

Dr. Gay E. Reinartz greets Kosana and baby Deidre through a window into the bonobo exhibit at the Milwaukee County Zoo.

Photo by Richard Brodzeller

The Stearns Family Apes of Africa building, which opened in 1992, was designed to facilitate the fission-fusion social organization followed by bonobos. Indeed, the acquisition of the bonobos from Wassenaar led to the $8 million fund-raising, planning and construction effort that resulted in the new building, according to Dr. Bruce Beehler, Deputy Zoo Director in charge of animal management and health. Fully half the structure is devoted to the bonobos, which enjoy an all-weather, skylit exhibit three stories high and filled with climbing structures, hiding places and a pool. Just as important is the layout of the off-exhibit area – featuring one outdoor enclosure (sadly, not easily accessible to visitors) and 10 indoor enclosures connected by a variety of passageways and chutes planned to accommodate a fission-fusion lifestyle. Some groups can easily bypass others (with whom they may not presently be getting along) and still arrive in the exhibit area or the outdoor enclosure.

Because bonobos are such social beings, a quarantine area was built within view of the rest of the indoor enclosures so newcomers could see and be seen before actually joining the group. Now that medical procedures are a standard part of bonobo life, further structural modifications are being made to allow for X-rays with a portable machine. Except for those times when health checks determine the placement of a bonobo in one enclosure or another, group dynamics determine who is where and with whom. There is a reason, said Dr. Victoria Clyde, one of the Zoo's veterinarians, why keepers are nicknamed "Gateways to the Forest."

Not that keepers get to have the final word about who goes where and with whom, Bell notes. If the bonobos like the grouping they hear the keepers describe, they line up together one after another, ready to transfer to the playroom or outdoors or the public exhibition area, she says. If they have other ideas, they line up in their own clusters.

Listening In

Photo by Richard Brodzeller

"They eavesdrop on our conversations, and they certainly recognize names and many other words they hear every day," she says. "It's a little like being in charge of a large and very interesting family."

Because bonobos' matriarchal society is so rich and complex, finding a place in the group is never automatic, even in welcoming Milwaukee, with its international reputation for taking in hard-to-place primates. Introducing the new kid on the block is different for every fresh face, according to Bell.

Tamia, an 8-year-old female from the Columbus Zoo, was sure of a warm reception when she came to Milwaukee late in 2004. She was born at the Columbus Zoo, with which Milwaukee has a long and

Tamia

mutually supportive relationship. At Columbus, the norms of bonobo society are observed. Moreover, Tamia was reared by her mother, not a human. Nuances of bonobo behavior come to her naturally. Her migration into the group at adolescence mirrors conduct in natural bonobo groups, where it is females who leave home for another part of the forest and males who stay with their mothers' group for life.

"She'll click with the girls easily," Bell predicted even before Tamia's mandatory quarantine period had ended. "We have a number of females around her age; so she will have lots of friends."

As a nubile youngster just entering sexual maturity, Tamia was an instant hit with the male population. She attracted the bonobo equivalent of wolf whistles when she took up residence in a separate but visible enclosure in the off-exhibit area in January 2005. However, Tamia won't be allowed to add motherhood to her attractions any time soon. She and 7-year-old Zanga Mokila, one of Laura's daughters, both are maturing early as a result of the enriched diet available to Zoo residents. In the wild, bonobo females don't bear young until they are about 14 or 15. Both Tamia and Zanga will be on birth control until they are physically larger and socially better prepared for motherhood.

The young female from Columbus immediately began to campaign for a patron among the power females who ensure that everybody else lives up to group standards. No sooner had Tamia been cleared from quarantine to join the group than she began grooming some of the matriarchs, a gesture of respect.

"She's charming," said the keeper.

ALL ABOUT EVE

Tamia also is ambitious. After ingratiating herself into the power chick clique, the group of females who set the rules and hand out discipline, she joined Linda in administering a painful rebuke to Brian, the 18-year-old who is jockeying to take over as dominant male from longtime leader Lody. At present Tamia is firmly enough ensconced among the most forceful females to get away with occasional disrespectful behavior around Maringa, who seems to be relinquishing her two-decade role as undisputed queen of Milwaukee's bonobos.

Not every newcomer has such an easy time as Tamia. The matriarchs heading Milwaukee's bonobo group are fiercely devoted to enforcing the community's code of conduct. Infractions often bring brutal retribution, even if they are unintentional.

A case in point is Viaje, a 27-year-old male from Mexico, whose name means "journey" in Spanish. Three years after arriving at the Milwaukee County Zoo, he was still working to win acceptance in 2004. As a wild-captured bonobo, he looms large in the Species Survival Plan. However, his 20 years in a private Mexican zoo was a study in abnormality. Viaje was stranded alone in a private zoo by

Photo by Mark Scheuber

Viaje

international regulations designed to protect human and animal health.

After mountains of paperwork and an epic obstacle course of a trip, Viaje arrived in November 2001 at the first large colony of bonobos he had experienced since his capture in 1980. The healthy newcomer was raring to make friends and breed with females, but he had no idea how to work and play well with others. An early encounter with the larger group was a disaster.

"The notion of rank (an essential feature of bonobo culture) was completely outside his understanding," said Bell. "The first time the whole group ate together, he took the best mango. If you are a low-ranking adult male, you simply don't do that."

Punishment was swift and painful. The group's power females – Maringa, Linda and Ana Neema – bit his hands until they bled.

"Did I plan it? No," said Bell. "Did it probably have to happen? Yes."

Zookeepers here are committed to enhancing the communal as well as the individual lives of those beings in their care. That means allowing the bonobo community to maintain and promote behavior as close as possible to that observed among their relatives in the wild.

"It may be primal," Bell said of the often furious response to rule-breaking. "In the wild, survival depends on group cooperation, and anybody who doesn't understand how the group operates just can't be there. In captivity, they can't chase off an offender; so they make sure that offenses aren't repeated."

CLUELESS FROM CUERNAVACA

Viaje

Photo by Mike Nepper

Viaje was not being boorish. He was clueless. A refugee from an era in zookeeping when social primates were kept with no more company than a mate, Viaje spent years with a female bonobo at a zoo near Cuernavaca. When the female died in 1993, possibly of tuberculosis, the zoo owner immediately tried to find another and larger community for the survivor. However, efforts were slowed and almost curtailed by strict international standards designed to prevent the spread of diseases such as TB, which affects most primates (including humans) and numerous other animals. With the cooperation of zoo owner Marcos Orteiza, the SSP sent a team of health experts to examine Viaje using the latest techniques to determine whether he had been infected with the disease. Veterinarians from the Audubon Park Zoo and Tulane University in New Orleans facilitated the time-sensitive import and testing of blood samples and thoracic washes tested by labs in the United States that had to be approved by the U.S. Center for Disease Control (CDC). Finally, it was determined that Viaje did not have active tuberculosis. Dr. Reinartz set about finding him a home. There were no immediate takers. Eventually, the Milwaukee County Zoo – with its long track record of successfully socializing problem animals – extended a welcome. The complicated tests and medical examinations had to be repeated. There were long consultations with the CDC to develop an importation strategy that would minimize health risks to humans and Viaje.

Only heroic persistence on the part of Orteiza, SSP committee members and staff at the Milwaukee County Zoo, including Jan Rafert, Milwaukee's curator of primates and small mammals, enabled Viaje's eventual arrival in Milwaukee. The process took five years. Since Viaje could offer some important genetic diversity, incorporating him into one of the planet's healthiest captive-breeding groups was a matter of no little importance. During Viaje's years of waiting for a new home, Orteiza became a substitute companion.

For years in isolation, Viaje was alpha male by default. Then Viaje arrived at the Milwaukee County Zoo, where there was an established social order. Suddenly he ran head on into bonobo etiquette in which dining protocols were as complicated as anything devised by 19th-century Victorian England. As a healthy, sexually mature male, Viaje (judging by his behavior) saw himself as a prime stud muffin and the group as a splendid array of potential sexual partners. To the group (judging by behavior), he appeared as a small male displaying the kind of disregard for community customs that would be tolerated only by the best beloved of their infants. He was showing himself to be untutored, uncouth and unfit. To win acceptance, Viaje had to discover that there were rules and that there were unpleasant consequences for failing to obey them.

Once the group's ruling females drew his attention to his inadequacies, he was ready to learn. Then Bell and other keepers could establish safe environments in which he could figure out how to mend his manners. In a separate enclosure adjacent to the main group during meals, Viaje could pick up slowly on the subtle cues that reveal dominance and privilege and on the behaviors that constitute courtesy. It also allowed him to be available for the kind of sexual activity that is a primary form of communication among bonobos.

For a long time, Bell placed him only with one of the group's passive females, gentle Kosana or the elderly, blind Kitty. Then, as his social confidence began to return, there would be an occasional, brief addition of one of the dominant females, often at a time when she was certain to be distracted by other activity. Viaje still was separated from the group during eating, which has complex codes of behavior. By concentrating on one thing at a time, he had a better chance of figuring out the rules. "We still don't try to combine social skills with eating skills, but he's learning," said Bell.

Bell considers him still clueless about the mercurial alliances that accompany the present changing of the guard in bonobo leadership. Viaje, however, recognizes a powerful patron when he sees him. On those occasions when he is allowed to be the humble sidekick of Brian, one of the leading contenders for alpha male status, Viaje enjoys the prerogatives of sexual contact and control of territory that come with the temporary friendship. As for the pitfalls of dealing with the group's ruling females, he is learning well enough to be favored by attention from Maringa, who at the time they mated was still the major player in group politics. Her latest offspring – Faith, born February 19, 2005 – was sired by Viaje.

Photo by Mark Scheuber

Faith, Viaje's offspring

From Chaos To Communication

Ana Neema with baby Gilda

CHAPTER 4.

From Chaos to Communication

When Barbara Bell began work as a keeper at the Milwaukee County Zoo in 1989, almost everybody on staff agreed on two things about the six bonobos then in residence: From a conservation standpoint, they were among the rarest and most important of the Zoo's treasures. From a practical standpoint, they were trouble. For sheer unpleasantness, the bonobos were unrivaled. In the off-exhibit areas where bonobos sleep and where contact with keepers is closest, the primates urinated on their caregivers with the accuracy born of malicious intent. When they weren't urinating, they threw feces. Sometimes they did both. They screamed in evident rage. Nor did they confine their anger to shrieks and insults. Not even the lions or elephants were more dangerous to approach. One keeper lost a finger to a bite.

Barbara Bell in front of the Zoo's bonobo exhibit.

Photo by Richard Brodzeller

Bell was fascinated.

Because she was a longtime friend of Sam LaMalfa, then primate area supervisor, Bell had seen the first bonobos shortly after they arrived at the Zoo in 1986, long before they went on exhibit.

"Their level of intensity drew me to them," she said.

In 1989, she had only occasional contact with primates. Her first assignment was to act as a rover, relieving other keepers on their days off or during sick days. She continued to be intrigued, however, by the bonobos – the talk of the Zoo staff for their amours toward each other and their aggression toward their caregivers.

"The situation was chaotic," Bell recalled.

Not even such a simple task as shifting a bonobo from one area to another could be accomplished without risk. Many times stronger than humans of the same size and weight, bonobos repeatedly tore apart the portals separating exhibit enclosures, eating areas and sleeping spaces. One year it cost the Zoo several thousand dollars just to replace damaged doors. Yet Bell's fascination simply grew. She asked to be assigned to primate duty, and LaMalfa gave her primary responsibility for the bonobos' care.

"It seemed to me that they were communicating," she said. "We were getting the message all right. They weren't happy. We just didn't know why. I was scared, overwhelmed and aware that I was in way over my head. I didn't know whether to laugh or cry. I knew that you couldn't manage bonobos as you did gorillas (which were long-time residents at the Zoo), but I didn't know what to do."

In the long run, one of the most effective avenues to understanding has proved to be empathy. Bell put herself in the place of the bonobos, whose intelligence she never questioned. What would make the most difference for her, if she were in their position? Bell, an avid reader with a lively mind that, by her own admission, "always needs a project," suspected that the answer, in part, was intellectual stimulation.

She had two immediate allies among her fellow keepers, Rebecca (Becky) Loehe (LAY) and Patricia (Trish) Khan. Khan is now area supervisor for primates. Loehe left the Zoo staff after a stint as head gorilla keeper. All three were members of a new generation of keepers who had backgrounds in natural science. Khan, who grew up caring for horses at a stable, had a degree in biology. Loehe, a dedicated animal lover, was pursuing a master's in physical anthropology. Bell brought a degree in environmental conservation to her new responsibilities as bonobo keeper. Loehe and Khan were relief keepers, but their daily assignments mattered far less than their enthusiasm about training bonobos. As soon as the new Stearns Family Apes of Africa building opened in May 1992, the three young keepers improvised short training sessions. Anything, they reasoned, had to be better than the current chaos.

"It was a mess," Khan said. "We really didn't know where to start. All we knew was that we had to do something."

Khan and Loehe would come over on their lunch hours from whatever area they were assigned. They started work on the mezzanine, high above the glassed-in area where the bonobos were on exhibit. It was out of the question to remove from public view, even for a little while, animals that had become such big attractions in the brand new building. At first the goal was really basic – name recognition for all the bonobos, not just a few. Once all the bonobos reliably responded to their names, the keepers tried getting them to focus their attention on symbols. Targets were homemade – geometric shapes attached to sticks sunk into concrete in buckets. Each woman tried to work with one bonobo at a time.

"We kept things short," Loehe recalled. "We'd bring a timer and set it for five minutes."

Meanwhile, they were seeking out every possible resource to improve their efforts. They talked with keepers at other zoos. (Over the years keepers in Milwaukee and colleagues at the Columbus Zoo, which now has a breeding group of 16 bonobos, have formed a particularly strong professional relationship.) They read books and scholarly papers. They sought out primate experts such as psychologist Dr. Sally Boysen, whose work with chimpanzees at the center she founded at Ohio State University has been featured on public television. They brainstormed with veterinarian Andrew Teare, then at the Milwaukee County Zoo and now clinical veterinarian at the Jacksonville Zoo. Teare had developed MedARKS (Medical Animal Record Keeping System), the software that became the zoo world's standard for medical record keeping.

Dr. Reinartz said: "It all came together at the same time. There were younger keepers who brought with them a different level of interest grounded in natural history education. They were face to face with these incredible animals. It was obvious that their daily job was not just a matter of cleaning out cages. But this transformation has not happened at every zoo. Barbara is just one of those people naturally gifted with animals. It really was her enthusiasm and her determination that created the basis for the training program."

There was good reason for the urgency the three keepers were bringing to their task. In May 2000, Bell and Khan presented a paper on their training program to a conference at the Brookfield (Illinois) Zoo on The Apes: Challenges for the 21st Century. Outlining their "Training in Multi-Task Medical Behaviors," they described the problems that precipitated their efforts:

"Providing the bonobos with medical care proved difficult due to the inability to separate animals for examination and their tremendous fear of the veterinarians. The bonobos also had an intense fear of anything novel, and variation in their routine brought immediate panic and aggressive behavior. Keeper and veterinary staff often relied upon trickery or bribery to manage the animals. Thus, the relationship between bonobo and caregiver was one built upon mistrust and fear. With the possible addition of more bonobos to the collection, it was necessary to develop a safer working environment and a less stressful existence for the bonobos. We decided that a positive reinforcement method of training would be implemented."

Bell and her colleagues consulted Shelley Ballmann, director of Oceans of Fun, Inc., the private seal and sea lion show at the Zoo, whose charges easily qualified as among the most cooperative on site. Ballmann is an expert in positive operant conditioning, a system of shaping behavior through voluntary participation and rewards. No one wanted bonobos to "perform," but everybody hoped that they might be persuaded to cooperate, at least in their own health care. Another source of good ideas was zoologist Dr. Mollie Bloomsmith, then at the M.D. Anderson Cancer

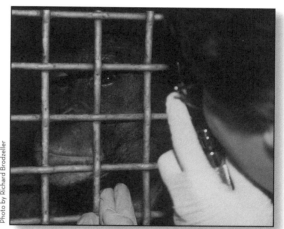

Photo by Richard Brodzeller

A bonobo gets close to Dr. Vickie Clyde so that she can examine its ear.

Research Center near Bastrop, Texas, now at Yerkes National Primate Research Center in Atlanta. Bell and Dr. Gay E. Reinartz enrolled in the workshop Bloomsmith was conducting in Bastrop in 1993. Khan and Loehe attended Dr. Bloomsmith's workshop the following year.

Dr. Reinartz recalled, "They had spent five years on a training program there (in Bastrop), shaping chimpanzee behaviors in order to improve husbandry procedures, and we were just in awe of what the chimps there were capable of."

The team's training efforts at the Milwaukee County Zoo began to take off in a big way. Bell proposed that the Zoo invest half the money already being spent to repair bonobo damage to their exhibit in hiring Ballmann as a consultant. With Ballmann's help and advice, the three young keepers – by this time all working full time in the primate area – applied what they had learned. Bonobos were free to participate or to avoid training completely. Armed with a tub of favorite food treats, the keepers worked with one bonobo at a time, using procedures taught by

Dr. Bloomsmith and reinforced by Ballmann. In the paper they presented at the Brookfield Zoo, Bell and Khan described their approach:

"Before beginning to train even the simplest of behaviors, keepers first had to establish a safe, positive working relationship with the animals. This was achieved by consistently rewarding desirable and nonaggressive behaviors. Undesirable behaviors were ignored and most were extinguished in four to six weeks."

Eye to Eye at Last

For the next three years, the three women worked together to develop the foundation of today's bonobo training program. "It wouldn't have happened without Becky and Trish," Bell said. Coming in early, staying late, they were involved in a constant round of trying new approaches, using what worked, tossing out what failed. All this took place while they cleaned the enclosures, chopped up and delivered food, and recorded daily behaviors for Zoo records. Little by little, keepers persuaded each bonobo to sit and focus, actually looking at the woman conducting the training and paying attention to the intended interaction between human and bonobo. Establishing eye contact was difficult because, among primates, direct eye contact often is viewed as a signal of hostile challenge. However, the bonobos found a way to distinguish between attention and aggression. Soon, training became not only a familiar activity, but a favorite spectator sport for the troop as well. Those not engaged in a training session would constitute an interested audience.

"We kept the sessions short and sweet with immediate rewards for focus and cooperation," said Bell. "The response was simply amazing. You could see the mental gears cranking. The level of happiness that they got from just learning things made me think of Helen Keller when she found out that objects she encountered had names. It became so obvious that we were tapping into a deep reservoir of personality. Finally, someone outside their group was sitting down and paying attention long enough to listen to them."

Equally startling were the differences in learning styles. Like their human counterparts, each bonobo was a unique learner. Lody, the alpha male, was fairly slow. Khan, who worked with him regularly, described a gestalt learning style with long periods of observation and study followed by performance that was rock solid. Lomako, his son, was just the opposite. Khan said he was always the star student, but sometimes his performance was inconsistent. Bell described Lomako as a learner who couldn't soak up information fast enough. By the end of the first year, the bonobos had learned to cooperate when keepers needed them to shift from one enclosure to another. They were no longer taking the place apart. Instead of hammering on the hardware, bonobos would sit in the transfer space and wait for the next door to open. Little by little, bonobos and their caregivers grew more at ease with each other. Keepers were educating themselves as well, discovering the distinct characteristics of their charges, adapting each session to meet particular needs and interests. Old ways of interaction were beginning to be replaced by something different – trust. On one of Bell's days off, Loehe and Khan were to discover in dramatic fashion just how vital that new relationship could be.

The day began normally enough. Loehe performed the complicated maneuvers that allow bonobos to leave their overnight quarters in the basement for their outdoor enclosure at the back of the building out of public view. She checked the outdoor facility to see that nothing dangerous had been introduced by accident overnight. She brought fruit and vegetables into the enclosure and placed it in several different areas. Then she went back through the tunnel leading into the keepers' area and began to open the series of doors that would give the bonobos access to the outdoors. After the last bonobo had left for the play yard, she shut the inside doors so she could start cleaning the overnight quarters. Then she went back outside through the human passageway to see that every bonobo was all right, the routine check. What she found was anything but ordinary. Lomako stood facing her, but he was outside the enclosure.

THE GREAT ESCAPE

Later Loehe and Khan discovered that the bonobos had systematically attacked the rivets attaching the heavy wire mesh to one support pole. They had opened a gap no larger than about four inches. Yet in the short time it took Loehe to shut the inner doors and go back outside, Lomako had wriggled through the opening and was now standing, with a big grin on his face, completely unconfined. Loehe immediately radioed for help.

Soon both veterinarians and almost a dozen keepers were assembled in the basement of the building. Khan and Loehe asked that all the backup help stay well out of view. Before resorting to tranquilizer darts or capture nets, the two keepers with whom the bonobos were most familiar would try persuasion. Loehe concentrated on maintaining eye contact with Lomako and keeping him reassured and calm. Khan had what her fellow keeper considered the more difficult task, of coaxing the adult bonobos – including Lody and Maringa, Lomako's parents and the group's dominant pair – back indoors. Lody had made his position as chief protector of the group unmistakably clear, and Maringa was loath to have any of the bonobos separated from each

Photo by Mike Nepper
Lomako

other, particularly Lomako from her. Loehe wondered to herself how Khan would be able to coax them to leave Lomako.

"Yet Maringa trusted me enough to come inside where she couldn't even see him, and Lody did, too," Khan said. "It was really a leap of faith for her. It would never have happened without all that training."

Once the adults were safely inside, Khan joined Loehe and literally scattered bananas in Lomako's path. To their surprise, he didn't take the bananas. He did not grasp Loehe's extended hand. He did follow her. Their progress was hesitant, slow and not without backtracking.

"I had gotten him all the way to the (human) door, when he looked up and saw all the guys with the nets and he retreated fast," Loehe said, "but we got them out of the way and got Lomako back on target."

Going in an unfamiliar series of doors was a strange experience for the young bonobo, but at last Loehe stood at the entrance to the overnight quarters and he, briefly, stood beside her. Then to her surprise and intense amusement, Lomako, like a jock in the shower room after a ball game, gave Loehe a light slap on her buttocks and ran happily into territory that was totally familiar.

Loehe said, "It was almost as if he had said, 'Hey, it was fun, but I'm home now.'"

Both she and Khan are convinced that only the trust that had developed over months of positive contact could have produced that result. Bell agreed.

"I don't want anybody to think that there was a magical transformation," said Bell. "There started to be some good times, but there were a lot of bad times, too, and there still are. Some bonobos still tried to bite. It wasn't all bunny hugs and warm fuzzies, but it gave keepers a chance to see something totally different in behavior."

Introducing a training program for bonobos was controversial, Bell said, because the very word "training" suggested circus animals and tricks instead of behaviors consistent with an animal's natural approach to life. Despite the controversy, the training program won the approval and support of Dr. Bruce Beehler, deputy Zoo director, and the rest of the Zoo hierarchy.

No Bonobos on Bicycles

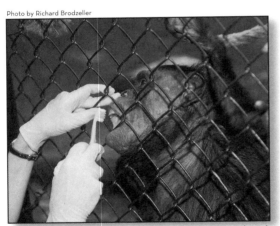

Photo by Richard Brodzeller

Laura loves having her teeth brushed. Bonobos have dental problems just as humans do.

Still, outside the Zoo, Bell recalls, "I caught a lot of flak. There was always that school of thinking that we were turning the bonobos into circus animals, training the apes to ride bicycles and jump through hoops of fire. (Critics) accused me of 'de-culturing' the animals, saying the bonobos would no longer be bonobos. What did make sense to everyone was training for medical procedures. In order to spare the animals from having to undergo anesthesia for a medical exam, people were happy to train them to help in their own medical care."

Loehe remembered vividly the fear that the bonobos used to show with the approach of a veterinarian. Training for medical procedures made life easier for both the veterinarians and the bonobos, she said. The new rapport was achieved just in time to make a dramatic difference in the life of one of the Zoo's original bonobos. Kidogo, who came to Milwaukee from Wassenaar, had returned to the Milwaukee County Zoo in November 1995 after a few years at Yerkes. He was diagnosed with severe cardiomyopathy. Dr. David Slosky, a cardiologist from St. Mary's Hospital in Milwaukee, was willing to consult on the bonobo's treatment. Could a training routine engage Kidogo's cooperation with the medical procedures needed to monitor his health? It could. Although Kidogo died of his heart condition in July 1996, the care he was able to receive eased the final months of his life.

A Shot in the Arm

One thing led to another. Five of the bonobos developed strep throat. The Zoo veterinarians needed keepers to teach a behavior that would allow throat swabs without anesthesia. A training program was devised to make it possible. As more training led to better cooperation, other questions arose. Could the vets weigh bonobos without anesthesia? Yes. Could the bonobos learn to permit injections? (The question had immediate importance. Linda had diabetes and needed regular insulin shots.) Once more, the answer was yes. Could they tolerate blood pressure cuffs? Yes. Could pregnant females cooperate with ultrasound examinations of their fetuses? Yes. By the time Bell and Khan presented their paper at the Brookfield Zoo in 2000, the training program had generated a wealth of baseline medical data on bonobos, largely in cardiology and obstetrics.

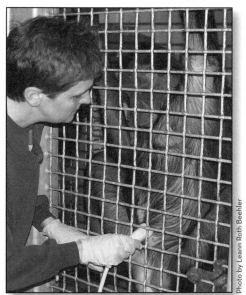

Photo by Leann Roth Beehler

Zookeeper Barbara Bell holds an ultrasound probe against bonobo Ana Neema's abdomen.

Leann Roth Beehler, who at the time owned Milwaukee's only multi-specialty mobile ultrasound company, volunteered the use of her equipment to provide an extra layer of prenatal care for the Zoo's expectant mothers, including the bonobos. Bell used her knowledge of group dynamics to ensure that all the troop's females would willingly participate. The first to be involved in an ultrasound examination was the troop's matriarch Maringa, the ultimate trend setter and role model. As a result of these efforts, the Zoo has an unprecedented library of bonobo ultrasonic images documenting normal and abnormal cardiac function and normal fetal development. Bell and Roth Beehler now are developing equipment and training procedures to collect blood pressure measurements.

What began as a program in operant conditioning has evolved far beyond an attempt to coax cooperation from difficult animals, Dr. Reinartz noted. It has led to a major change in the way zoos throughout North America monitor the health care of animals. Since she began serving as veterinarian advisor for the Bonobo Species Survival Plan in April 1998, Dr. Victoria (Vickie) Clyde, one of the Zoo's veterinarians, has used every possible opportunity to let keepers and her fellow veterinarians know the benefits of medical monitoring without anesthesia. Her work with bonobos here has led to numerous publications and presentations at conferences throughout the country. From a careful analysis of the studbooks that keep records of births and deaths of captive bonobos here and in Europe, Dr. Clyde discovered that 40% of adult bonobo deaths

Photo by Richard Brodzeller

Dr. Vickie Clyde, veterinarian

were due to heart problems. The last thing a vet wants to do is anesthetize an animal with heart problems, she said. Therefore, other veterinarians have been delighted to hear that there are ways to monitor animal heart functions without anesthesia.

She said: "I want people to understand that these procedures are possible if there is a commitment to operant conditioning. Keepers who are enthusiastic about a training program need the support of management to make it happen. Therefore, it's important to let the zoo directors and veterinarians know that there are genuine medical benefits. I also want to let people know that these training programs are fun and enjoyable, for the vet as well as for the animals.

Earning and Learning Trust

"I try to get across the point that the biggest benefit of operant conditioning is that it develops trust. Medical procedures are just an extension of that. When animals trust us, it decreases the stress they suffer, which is always a component of disease. They are more comfortable displaying signs of illness. We get a better assessment. To me, being able to do a physical exam or get swabs of their throat or take a blood sample is the icing on the cake. It's another way we can do work for their benefit."

Keepers, according to Dr. Clyde, are the key to the whole process.

She said, "It's Barb Bell's face lighting up when she introduces me by saying, 'Look who's here for training' that builds the bridge to the bonobos. We earn the trust because the keepers extend it to us."

The physical health benefits have been significant. Because throat swabs and blood samples are possible, doctors now are better informed about the causes of another major threat to bonobo longevity: respiratory infections. All captive collections of primates are subject to intermittent outbreaks of respiratory disease, she said. With information collected at Milwaukee and other zoos, veterinarians can document not only what infections the bonobos are getting, but also which are most likely to be dangerous.

Photo by Richard Brodzeller

Laura and Makanza touch Bell's hands during a voluntary operant conditioning session.

There also are major mental health benefits from the operant conditioning program, Dr. Clyde said.

"They enjoy it," she said. "Just like people enjoy going to school and learning something, training is a delight for bonobos. They are curious, social creatures that love new challenges."

As cooperation increased, so did the roster of experts outside the zoo staff who were willing to donate their services to treat bonobos. Dr. John Scheels, a Wauwatosa dentist, has been caring for animals with problem teeth at the Milwaukee County Zoo for 25 years and is internationally acclaimed for his dental

Bonobos: Encounters in Empathy

expertise with exotic animals. He was there when Kitty, the oldest member of the troop, needed several teeth pulled. Dr. Frank P. Begun, a urologist at Froedtert Memorial Lutheran Hospital, also has been a long-time member of Dr. Clyde's call list. When bonobo squabbles escalate to blood-letting, Dr. James R. Sanger, a plastic surgeon at Froedtert, sometimes steps in with a suture or two.

Even Milwaukee bonobos temporarily living elsewhere have received extraordinary care. Naomi, an 8-year-old female who was one of the original seven bonobos to come to Milwaukee, had been on breeding loan at Yerkes National Primate Research Center for five months in 1990, when she ate a plastic ball given to her as a part of enrichment activities. After two surgeries to remove plastic material from her bowels, Yerkes officials contacted Milwaukee with a request for euthanasia. Instead, the Zoo here requested a second opinion, hired a Lear jet and had the ailing bonobo flown to Froedtert for CAT scans. Unfortunately, Naomi died during the scan, but the incident shows the extent to which the Zoo and its medical allies have gone to protect the precious primates.

THE MILK OF HUMAN KINDNESS

With help from another kind of expert – lactation consultant Patricia Gima – Bell even found a way to facilitate the rearing of young Deidre. When this female bonobo was born March 4, 2003, she was smaller than most newborn bonobos. Her mother, Kosana, had no previous success in mothering her own babies. Deidre seemed to need more nutrients than she was getting at her mother's breast. Supplementing breast milk with formula delivered by a keeper would have led to a reduction in Kosana's milk production and would have gotten in the way of the nurturing mother-infant bond reinforced by suckling. Instead, Bell followed Gima's advice to enrich the mother's diet. Within a month, Kosana's milk production had increased, the baby was plumper, and Kosana was becoming an experienced mother.

Photo by Mark Scheuber

Deidre responded to her mother's enriched diet and put on weight.

"I'm always amazed at the people who offer help when we need something for the bonobos," said Bell.

In 1997, Dr. Harry Prosen, then chairman of the psychiatry department of the Medical College of Wisconsin (in Milwaukee), joined the group of experts on call for the bonobos. He was asked to consult with Bell on a severely disturbed newcomer to the group, Brian. The psychiatrist was immediately struck by the strength of the relationship he observed between bonobos and their keepers. Dr. Prosen had spent a lifetime treating patients with what he calls "an empathy deficit." At the Zoo, he noticed instead "an empathetic connection between keepers and those in their care. It was fascinating."

Daily life for the bonobos had been revolutionized in response to the cues – first given in anger and later in cooperation – given by the primates themselves. No longer distracted by the constant need for defensive strategies, keepers could pay close attention to the preferences, natural groupings and emotional comfort of their charges. There was a new level of rapport between keepers and bonobos that allowed recognition of individual differences and preferences and that enhanced the development of bonobo personalities.

Photo by Mark Scheuber

Maringa grooms Lody while Faith lounges in his arms.

Patterns of group behavior ingrained in the four wild-caught members of the troop then in residence – Kitty, Maringa, Lody and Linda – were reasserting themselves. Females were dominant, supported by Lody the alpha male. Power females took on the responsibility of physical reprimands, which were often swift and savage. Sexual interaction, in all possible combinations, was used more often to ease tensions and establish connections between individuals. Mothers suckled their infants for years. Male political power in the group was linked to the mother's position. With the comfort level rising among the bonobos, Bell's problem primates were turning out to be an astonishingly normal group, captivity notwithstanding.

"We couldn't do anything about their captivity," said Bell, noting that the wild-captured core group had left Zaire at a time when animals were widely traded on the international market. "We could do something about their living conditions; so we concentrated on that."

JUST JOKING

Through the regular interaction that training provided, keepers began to understand some of the nuances of bonobo life, among them the robust sense of humor shared by their intelligent charges. Bell recounted a hilarious encounter between a keeper and Lody, who once took his leadership responsibilities so seriously that he rarely smiled. A keeper was asking him to extend his left hand over and over again. Time after time, he extended his right hand.

Photo by Richard Brodzeller

Bell said, "I noticed this little quirk at the corner of his mouth, the kind of thing we do when we're trying to repress a smile."

The keeper sternly responded to the latest "mistake" by saying in a strong, firm voice, "No, Lody, your *left* hand."

Bell said the large male, his face still a mask of serious concentration, reached behind him with his left hand and in a leisurely way scratched his rectum. Then, still with a completely straight face, he extended to the keeper his liberally anointed left hand.

"We all roared with laughter," Bell said, "and he rolled on the floor laughing. He thought it was so funny. There was no doubt that he had made a joke."

She readily credits the troop leaders with making the transformation possible. Maringa and Lody – the unchallenged royal couple until late in 2005 – "knew how to be bonobos and they set the example for everybody else."

Lody, the wise elder and sometimes jokester of Milwaukee County Zoo's bonobos.

Bonobos: Encounters in Empathy

Dr. Clyde still defers to Lody as alpha male. She had been involved with training sessions for two or three years before he stopped charging her. Every time she appeared, he would rush toward her in a threatening posture to proclaim his role as the group's primary protector. She got the message. Before beginning a medical procedure, she approaches him first. It's simply a matter of courtesy to a leader, she explained.

"The females enforce the social rules, but the males still act as arbiters," she said. "If Lody is calm with a procedure, everybody else is. It takes no extra time for me to acknowledge his position in the group."

After years of scrambling for information and understanding, Bell now has found herself in the consultant's role as she gives presentations to zoo meetings in the United States, offers training to primate keepers from institutions as prestigious as Planckendael Animal Park in Antwerp, and answers e-mail queries from around the world. There is even evidence that there may be a change in practice among some zoos which, in the interest of infant survival, previously had removed bonobos from their mothers at birth to rear the newborns in a nursery with human care. The Bonobo Species Survival Plan guidelines for animal care clearly recommend that bonobos be reared by their mothers. Staffs at some zoos, however, have been so fearful that the infants might suffer in the care of inexperienced mothers that they repeatedly have applied for (and received) exemptions from the rule to allow keepers to rear the offspring.

Photo by Richard Brodzeller

Ana Neema and Gilda

Natural nurturing patterns that have been so successful for the seven surviving bonobos born at Milwaukee's zoo have begun to be noticed. (A female was stillborn to Lannie, one of the original seven, who died eight days after the birth in 1987. Eliya, 7, one of Maringa's daughters, died of pneumonia in 1998. Yatole, a female, died of congenital defects at age 10 months in 1996. All the other offspring thrived.) In Milwaukee and other zoos that have closely collaborated on bonobo-management strategies, zookeepers have followed both SSP guidelines and the bonobos' lead by leaving newborns with their mothers. The practice has yielded mother-reared youngsters as socially robust as they are physically healthy. Dr. Clyde said the differences between mother-reared and keeper-reared bonobos have been dramatic enough to convince the larger zoo community that SSP guidelines are even more important than originally understood.

WORLDWIDE CELEBRITY

Only two decades after being introduced to zoogoers who often asked, "What's a bonobo?" Milwaukee's bonobos have become nationally and internationally famous. They have been featured in the Sunday Times of London, *Der Spiegel* (a top German news magazine), the New York Times science section and on the Canadian Broadcasting Corporation and National Public Radio.

In 2006 National Geographic arranged to air a Pioneer Productions television special on mammalian fetal development called "In the Womb – Mammals. " The documentary includes photographs and data from the extensive record of fetal sonograms of seven Milwaukee bonobo pregnancies. In yet another research coup for the Zoo, the sonograms and fetal measurements – ranging from "grain of rice" size after conception to a fully developed infant a few hours before birth – have been painstakingly assembled and correlated by Roth Beehler. She also donated the ultrasound machine that made possible the data collection.

To Bell, however, the most satisfying aspect of the ongoing journey toward a culturally authentic bonobo community is the change in the primates themselves. They treat the Zoo as a home rather than a cage. When something is out of place, they notice and act. For example, in August 2005, Lomako found a bucket and brush left by a keeper in the playroom. Picking up the bucket in one hand and the brush in the other, he walked upright trying to get Bell's attention. When his movement failed to attract her eye, he knocked on the shift door, gave her the scrub brush and took the bucket to Kitty's pen. After the keeper opened the shift gate, he gently placed the bucket inside.

The primates take pleasure in Wisconsin, even during winter weather so different from their equatorial homeland. When Milwaukee received an early snow November 23, 2005, the bonobos greeted it as enthusiastically as a human child with a new sled. They rolled, danced and jumped in the snow and had lots of snowball fights.

Bell said: "There is a level of joy that is present because of the social environment. They enjoy the educational activities, but the best mental stimulation for them is the interaction they have with their peers. Their lives are rich with drama, emotion, intergenerational affection, play, argument, love.

"Even a nasty fight can be a healthy part of their lives together. They're like humans in that regard. Everybody gathers around to watch. The fights don't last very long. Vicious squabbles may result in injury, but they flare up and then it's over.

"Sure, there is posturing, dramatic displays, but for the most part, I hear laughter and play. There's a lot of giggling, teasing. Their lives are spicy. Something is going right."

Encounters in Empathy

Laura and baby Claudine

CHAPTER 5.

Encounters in Empathy

When psychiatrist Harry Prosen arrived at the Zoo to treat his first non-human patient, a bonobo, he didn't expect that he also would encounter his first non-human colleague. Yet he considers the group's alpha male, Lody, to be not only a fellow healer, but also a prime example of empathetic behavior. On a regular basis, Dr. Prosen visits the off-exhibit bonobo enclosures to research empathy, a subject that has engaged him for a lifetime. Often Lody is the focus of his attention. After a half-century as a psychotherapist, part of it as chairman of the psychiatry department at the Medical College of Wisconsin, Dr. Prosen is discovering new insights from his observations of humanity's closest relatives, the bonobos. The psychiatrist was introduced to the rare primates in 1997. He was consulted by Zoo staffers desperate to ease the pain of a newly arrived animal named Brian. Since bonobos along with common chimpanzees share 98.4% of human DNA, keeper Barbara Bell reasoned that there might be psychiatric similarities as well.

"It was worth a try," agreed primate curator Jan Rafert. "With Brian, nothing else was successful."

The 8-year-old Brian had just been transferred from a scientific research facility to Milwaukee at the behest of the Bonobo Species Survival Plan committee and its coordinator, Dr. Gay E. Reinartz, who is also the Zoological Society's conservation coordinator. In 1997 there were only 133 bonobos in all the world's zoos. By 2006, after years of careful breeding, the captive population still numbered only 160. No one knows how many individuals survive in their sole natural homeland, the Democratic Republic of Congo, where poaching, rain-forest destruction, economic collapse and civil war all have taken their toll on the species. In 1997, everyone concerned about bonobos agreed that the population could not afford to lose a single individual, however tormented.

Photo by Mark Scheuber

Brian

ARRIVING IN ANGUISH

Brian's distress was unmistakable. He communicated his anguish with everything he did – shrieking, pacing in a tight circle, clapping his hands, inducing vomiting, mutilating himself, rarely eating, never sleeping. Bell was haunted by the extreme nature of his suffering. The Milwaukee County Zoo had built a reputation for being able to accommodate problem animals largely because of the mental stability provided by the dominant pair, Lody and Maringa. Accustomed to new arrivals, the staff had devised multiple techniques for establishing rapport with newcomers. With Brian, Rafert recalled, keepers tried every one of those stratagems. With Brian, none of it worked.

Dr. Reinartz called Dr. Prosen. Like a series of other experts consulted by the Zoo, the psychotherapist was intrigued by the challenge. Dr. Prosen brought to his unusual new client the same methodical approach he would have used with a human patient, beginning with a careful perusal of a written case history from the research facility where Brian was born and spent his first eight years. One particular event caught Dr. Prosen's attention. Brian's infancy had diverged from the normal pattern for bonobos in which mother and baby are in constant contact. For reasons that remain unexplained, young Brian was confined for a time in a relatively small area with no companion but his father.

Following the procedures of a lifetime, Dr. Prosen convened a case conference with Zoo veterinarians, animal curators, keepers and Shelley Ballmann, director of the private seal and sea lion show at the Zoo who had helped keepers devise training and enrichment activities for most of the animals. Dr. Prosen was immediately taken with the perceptive observations of those who provided daily care to Zoo animals of all species. Zoo staffers, in turn, were amazed that Dr. Prosen had treated human patients with behaviors similar to those of the agonized bonobo.

SAFE AND SAME FOR BRIAN

A clearer picture of Brian began to emerge. His self-destructive behavior, Dr. Prosen reasoned, rose from desperate, but futile, attempts to calm himself in the midst of extreme anxiety. The psychiatrist worked out an alternative, less dangerous path to peace. Dr. Prosen prescribed surrounding Brian with a world that was absolutely predictable. Bell and the rest of the staff fit into their busy schedules time to provide a fixed, dependable routine for Brian that would have bored his normal fellow bonobos into screaming fits of frustration. His meals were at the same times and in the same places, always. His quiet time followed his lunch, always. Keepers approached him using the same gestures, the

Photo by Mike Nepper

Unvarying mealtimes were part of Brian's therapy.

same tone of voice, the same words of praise, always. New experiences were introduced in slow motion. For example, a new toy might be placed some distance away from Brian's enclosure but in his sight. Later it would be moved within touching distance. Finally, it would be placed inside the enclosure.

In essence, Dr. Prosen said, keepers created a therapeutic environment in which healing could take place. Bell credits her fellow keepers with making the treatment possible, despite the tough workload they already faced. Enclosures, for example, have to be thoroughly hosed out twice a day. Zookeeper Mark Scheuber picked up the extra duties without a murmur. Patricia "Trish" Khan, primate area supervisor, pitched in to keep training on schedule. Claire Richard, principal gorilla keeper, also lent a hand in between feeding, cleaning and training chores with her own massive and complex charges.

"Bonobos are a heck of a lot of work to begin with," Bell said, "and this just meant more work for everybody."

Companions were introduced to Brian with equal care. Here, the psychiatrist became even more captivated by the bonobos. In the interactions of some other bonobos with Brian, Dr. Prosen began to recognize dramatic examples of empathy — both the empathetic connection exhibited by psychologically healthy humans and the empathetic deficits shown by those with psychological difficulties.

Helping Miss Kitty

Because Brian's behavior was so aberrant, most members of the group either avoided or repudiated him. His behavior toward females was not that of a mature male, but of an infant. When he tried to nurse from one female, Laura, she merely rejected him. Linda, one of the dominant females, who frequently acts as an enforcer of group norms, was much more emphatic. She bit him bloody. However, Brian also found acceptance from his fellows. Aged Kitty, now 57, was already blind and deaf when she and Brian were introduced. Yet she often engaged in grooming him, an activity he seemed to find calming. In turn, Brian was seen gently helping Kitty into the outdoor area.

Photo by Mark Scheuber

Kitty likes to sit in the sun.

Keeper notes for July 12, 1998, record the incident: "It is a milestone for Brian to be the helper, instead of the one needing assistance."

Other forward steps followed, but not without mishap. Generally, bonobos adopt a make-love-not-war approach to living with others. In contrast with the male-oriented societies of common chimpanzees, bonobos form communities around matriarchal groups. Although the more forceful females also administer painful physical discipline, the group's matriarchs rule mainly by manipulation and the granting or withholding of sexual favors, Bell said. Sexual contact of all kinds is used to defuse tension, to resolve conflict, to aid reconciliation and to strengthen bonds among members of the group. However, sexual behavior codes and dominance patterns are complex. Often an outsider gets it wrong, and transgressions – usually invisible to human observers – are swiftly and severely punished. For a long time, Brian was on the receiving end of painful rebukes.

In September 1998, Linda delivered bites that took weeks to heal. Yet just two days later, Lomako, one of the group's senior males, spent the day pampering, cuddling and grooming the wounded Brian, "a much needed mental health boost," a keeper noted. Dr. Prosen has been particularly impressed by the attention given Brian by Lody, the group's dominant male. When Brian seemed too panicky to move, Lody often would take his hand and walk him into a different area such as the playroom or the outdoor yard. It was in Lody's company that Brian first sat down and ate with a group, a "very big deal for him," recorded in animal-management notes August 30, 1999. Regularly, Lody would postpone his own meals to sit with and comfort Brian when the younger bonobo was having difficulties.

Brian received just the opposite treatment from Murph, a younger male bonobo acting out an extremely annoying adolescence by teasing almost every other member of the group. Distributing favorite foods hidden in some way is a frequent practice to provide variety in the lives of these intelligent primates. On one occasion treats were concealed in mailing tubes. Murph immediately grabbed Brian's as well as his own, whereupon Lody stuffed his food back into his own mailing tube and presented it to Brian.

Requiem for a Fallen Comrade

In Lody, Dr. Prosen said, the presence of weakness seems to activate his own compassionate strength. One of the Milwaukee County Zoo's original bonobos, Kidogo, returned to the Zoo in 1995 after five years on breeding loan to Yerkes National Primate Research Center in Georgia. The 21-year-old male was suffering from congestive heart failure. Kidogo and Lody both had been at Dierenpark Wassenaar and together made the journey from the Netherlands to Milwaukee in 1986. When Kidogo returned to Milwaukee from Yerkes, Lody became a constant companion to the fragile male. Often Lody walked with his arms wrapped tightly around Kidogo. He carried or assisted the ailing ape in climbing ladders in the exhibit and would bring the invalid the most coveted food items. When Kidogo died in 1996, nine months after his arrival, his death was mourned by his long-time friend. Lody lay on a sleeping platform and refused to eat for almost a week. He remained emotionally withdrawn for another two weeks.

Nor were adult bonobos the only objects of Lody's concern. On two occasions he has assumed the role of surrogate father – rocking, cradling and playing with the baby in his care, often skipping a meal to attend to the infant's needs. When 2-year-old, keeper-reared Zuri arrived in Milwaukee, Lody eased the newcomer's introduction to the rest of the group. On their first encounter, the dominant male scooped up the youngster for a few hugs and then calmly stayed near without forcing his presence on the infant. Keepers at the facility where the little male was born revealed that Zuri had never been carried or held by another bonobo before coming to Milwaukee. Instead, the infant usually clutched a security blanket, a habit that did not escape Lody's attention. One day when Lody and Zuri were leaving one area for another, the dominant male carefully collected the blanket before the two moved into the new location.

Hugs and Giggles

On occasion, younger male bonobos, including the troubled Brian, have seemed to model their own behavior after that of the alpha male. Only two months after Zuri arrived in Milwaukee, Brian followed Lody's example in regard to the infant. As gently as he had observed Lody doing, Brian held the infant all day long. Ten days later, he was playing non-stop with the youngster, and keepers noted, "Brian is exhibiting new play behavior. Zuri is a huge positive in Brian's life." When Linda eventually took over maternal duties, assuming primary care of the infant (normally bonobo babies stay close to their mothers for the first four years), Lody and Brian withdrew to more peripheral roles.

Photo by Mark Scheuber

Brian plays with Zanga Mokila. He learned this behavior from Lody.

Bonobos: Encounters in Empathy

Lody is patient and respectful with Kitty, elderly and infirm at age 57. With only her sense of touch and smell to guide her, blind, deaf Kitty easily becomes disoriented and confused. Lody gently leads her to her favorite sunny spot in the outdoor exhibit area. When it is time to return indoors, Lody usually wakes her and walks her inside. If the elder bonobo is having one of her frequent epileptic episodes, he often refuses to leave her.

When Viaje, usually the lowest ranking male in the group, arrived at the Zoo in 2001, Dr. Prosen and Bell noticed that Lody almost immediately appeared to identify the newcomer (who had lived alone for almost 10 years) as a "special needs individual." Viaje, who hadn't had to share food for a decade and who had no notion of his low rank in the group, was severely reprimanded by the power females for eating out of turn. Mealtimes became confusing and difficult periods for the new kid on the block, whose actual age was 21 but whose social demeanor was that of a juvenile. Bell observed that on many occasions Lody sat with Viaje and waited with him for a chance to join the shared meal.

Photo by Richard Brodzeller

Viaje

Such dramatic examples as these argue forcefully that bonobos are capable of empathy, according to Dr. Prosen and Bell. Bell cites almost a decade of observing the primates' daily interactions to confirm the findings of behavioral experiments conducted by Dr. Frans B.M. de Waal, director of the Living Links Center at Yerkes in Atlanta. Bell and Dr. Prosen further suggest that, at least in Lody's case, there is evidence of wisdom, in the Aristotelian sense: the ability to see life in all its aspects and to act in a way that benefits others. In the draft of a paper on which Dr. Prosen and Bell are collaborating, they note that Lody has incorporated empathy into his leadership style as alpha male, and that this quality has benefited the group in enormously important ways. They write:

"We became curious as to how this empathy develops and whether it is indigenous to all of the bonobos. As we examined the process, we became more and more impressed with its significance to the development of the group."

Specifically, Dr. Prosen and Bell noted that empathy kept members close to each other and helped establish constructive relationships between individuals. Reiterating his observation when he was introduced to the bonobos and their keepers, Dr. Prosen pointed out the empathetic understanding that had grown up between the humans and the primates in their care. In practice, he said, it was empathy that assisted keepers in understanding individual bonobos and group dynamics. Empathy allowed them to notice when tensions were high and to identify sources of conflict. Empathy led keepers to work out strategies to assist the bonobos in resolving the tensions and defusing conflict in ways that allowed individuals to develop and mature in as healthy a manner as possible.

ALTRUISM AND ADOLESCENCE

Dr. Prosen says his colleagues in the psychiatric community have been intrigued by the description of how "Lody's empathetic behaviors and ability to use good judgment in parenting skills, discipline and, in many instances, the demonstration of altruistic behaviors have had a powerful impact on the development of the juvenile males in the bonobo group."

Certainly, he said, there is evidence from the work of evolutionary biologists to suggest that it was the distant ancestors of bonobos that brought empathy into the evolutionary process. Through their observation of bonobos here and their study of each individual's history, Dr. Prosen and Bell have concluded that certain life experiences are necessary for a bonobo to develop into an empathetic and wise adult. Although Lody and Maringa both were captured as very young juveniles, they were at least 18 months and probably 2 years old before they were sold in Kinshasa to the sailors who brought them to Amsterdam.

"That means they had been breast-fed and raised by their mothers for a crucial period of time," said Bell. "If they had been younger than that, they never would have survived the voyage with only or mainly solid food to sustain them."

When Jan and Hanneke Louwman bought the young bonobos, which otherwise would have been sold elsewhere, the Dutch couple provided a nurturing environment without imposing human expectations on the juveniles. That nurturing care was obviously remembered by Lody and Maringa years later when the private zoo owners visited the Milwaukee County Zoo. When the Louwmans spoke Dutch to Maringa and Lody, the bonobos

Photo by Mark Scheuber

Lody grooms Maringa.

instantly recognized their former caregivers, Bell said. Rather than the appearance of the humans, much altered by the intervening years, she said, it was the language and the voices that seemed to trigger recognition. She said the two bonobos leaped into the air, hugged, hooted loudly and generally acted as though they were so full of emotion that they couldn't contain themselves.

THE LODY LOOK

Photo by Mark Scheuber

Lody, Maringa and Faith

Dr. Prosen makes the case that the nurturing care Lody received (and remembered) constituted a key element in the bonobo's maturing into an empathetic and wise male. As alpha, Lody has involved himself actively in the guidance of five younger males: his son Lomako, 23; the relative newcomer Viaje, 27; the troubled Brian, 18; the impudent Murph, 17; and the orphaned Makanza, 12. All five now are engaged in a series of confrontations and dominance games that ultimately will decide who follows the now-ailing Lody as the group's powerhouse male. Dr. Prosen and Bell have watched Lody rule by stern glance (they call it the Lody look) and threatening posture, getting physical only when necessary. Even Murph, who still acts the brat and with whom Lody frequently has had to use force, has shown signs that the lessons are received, Dr. Prosen said.

Bonobos: Encounters in Empathy

Sometimes, Lody allows others to administer needed discipline. Although he repeatedly had scolded or struck Lomako when his son played roughly with females or teased infants, recently Lody simply ignored Lomako's cries for help while the power females bit and hit the younger male. Afterward, Lody groomed Lomako for hours while the offender healed. Lomako evidently got the message. He now is respectful of the females and gentle with the youngsters. For a while in 2005, in a series of events that astonished observers for its Byzantine twists and turns, Brian seemed to be on course to depose Lody as alpha male.

"They still get along fine," says Bell, "but their roles are reversed. Brian eats first and Lody second. Frankly, I think Lody just doesn't want to be the leader any more."

As early as 2002, Lody began allowing Brian more and more importance. When Brian assumed leadership of a group of 13 one day, Lody sat back and let him organize to his heart's content. By spring of that year, Brian ranked second only to Lody in status, and swaggered around puffed up and full of himself, almost as though he knew that his personal psychological saga had been headline news in publications around the world.

Chastised by the same females who used to terrify him, Brian was beginning to hold his own. Bell noted that the former social misfit had learned the proper reconciliation gestures to win his way back into the good graces of even the female heavyweights – Maringa, Linda and Ana Neema – who still beat him up. She saw Brian offering respectful grooming and extending his hand, palm up in a begging posture that low-ranking Viaje uses to ask whether he can join in a meal. On another occasion, he crawled on his belly toward Lody for permission to rejoin the group.

A buff and handsome Brian (who has gained 40 pounds as he has matured) has cultivated the tolerance and even the affection of the ruling females. He has formed temporary alliances with both Viaje (who is just happy for a friend) and Lomako (who usually seems uninterested in establishing a dominant role for himself although he has been considered heir presumptive as the son of Lody and Maringa.) Brian has teamed up with Ana Neema to run Lomako out of the group on

Photo by Mike Nepper

Lomako, son of Lody and Maringa, is the logical heir to the role of dominant male, but he often seems uninterested in the role.

occasion. Grouped with Viaje and Laura, Brian wouldn't allow Viaje to breed her, Bell says. Challenged unexpectedly by Lomako, Brian delivered an unexpected but decisive beating.

Lately, Bell says, Brian has abandoned "the road rage approach." Whether the young pretender to the throne actually has picked up the lessons of empathy from Lody is subject to question, however. Dr. Prosen cautions that Brian might be one of those individuals with a real empathetic deficit. Some of his kinder behaviors may be no more than mimicry. Yet both Dr. Prosen and Bell say Brian has genuine admiration for Kitty and treats her with tenderness. Likewise, Brian is always gentle with the infants. Nothing makes him happier than playing with babies, Bell says.

A Rogue Wave in a Calm Sea

"He was a rogue wave in a calm sea," Bell notes. "I think he has quit fighting the system that is well established in this group." So far the real powers of the group – the dominant females – have accepted the idea of a new leader, whoever that turns out to be. "But there are checks and balances built into this group," says Bell. "Nobody can get too far out of line with the power chicks in charge."

Photo by Richard Brodzeller

Also, Brian can expect continued competition perhaps from the apolitical Lomako and certainly from the irrepressible Murph, who shows genuine ambition to be No. 1. Although Murph is more than a year younger than Brian, he has one advantage that may, in the long run, outweigh Brian's larger size. Murph's mother is Laura, one of the group's divas, and among bonobos, power usually descends from the female line.

Kosana's daughter, Deidre, looks like she's showing her female power.

Sensing Something Extra

After years of observing Milwaukee's bonobo community, Dr. Prosen and Bell have concluded that there is more to these fellow primates than just a high degree of intelligence. They are not the only ones to sense that extra dimension. Dr. Linda Cieslik left the Zoo staff in 1993 to pursue a career in education. In the late 1980s and early 1990s, when she was a doctoral student, she worked as a primate keeper in charge of the orangutans and Japanese macaques and as relief keeper for the bonobos when they first arrived at the Zoo. She was on duty at the Zoo's old primate building Dec. 17, 1990, when Maringa gave birth to Eliya, the first bonobo born in Milwaukee. Maringa was squatting as the other bonobos stood around her.

"I was going to call Sam (LaMalfa, the keeper in charge of the bonobos)," Dr. Cieslik recalled. "I turned away for a moment and when I looked back, there was this baby. All the other bonobos were gathered around looking very interested. They were pretty low-key too, cautiously touching Maringa and baby and keeping their voices down."

Not all her interactions with the bonobos were as benign. One day in 1989 when Dr. Cieslik was alone on duty, she was distributing vitamins to the group. Lody reached out suddenly and grabbed her under the left armpit. Turning toward him while she tried to loosen his grip, she repeated in a calm, even tone, "Lody, no." She succeeded in pulling her arm away until he was grasping only the last three fingers of her left hand. Her thumb and index finger were sticking out. The passage between the cages where she stood was too narrow for her to pull back completely out of his reach. Lody put her index finger in his mouth and bit down. As they both heard a crunching sound, he looked up, surprised, and she jerked her hand away. Although the bitten portion of her finger was recovered and taken to the hospital along with her, surgeons were not able to reattach it. After three days in the hospital, Dr. Cieslik decided against taking more time off. She returned to work and went right back to the bonobos.

When Lody approached the window, she raised her bandaged left hand and said, "Lody, my man, do you know what you did? He looked me in the eye and went to the back of the exhibit and sat there, with his arms wrapped around himself and his head down. I think he felt bad. I bear him no malice. Lody is a pal of mine. It just happened. I don't think it's his fault."

Nevertheless, Dr. Cieslik said she was deeply moved by Lody's behavior at the time of her return to work. She was even more impressed by an encounter years later. It was on July 17, 2004, on one of her infrequent trips to the Zoo. She was in the public area with Barbara Bell, outside the windows overlooking the bonobos' large indoor enclosure. The building was full of visitors. Dr. Cieslik was just one of many spectators.

"He could have just tuned me out," she said. "Instead he came running right over to the window. My hands were behind and below the railing where he couldn't see them, but he kept looking left, looking left."

Finally, she raised her left hand in greeting. He looked directly at the hand and then at Dr. Cieslik and then at the hand again.

"He knew," she said. "That was a moment when you know there's more there than a beast, something more than what we understand."

Dr. Prosen raises the obvious question: "Are we anthropomorphizing? That's possible, but all science is a matter of interpreting new things we are observing in the context of what we already know."

Days of Their Lives

Kosana (left), infant Deidre, and Tamia

CHAPTER 6.

Days of Their Lives

No one who has spent time paying attention to bonobos can fail to notice their keen intelligence. Nor does any close observer doubt their sense of theater. Like human actors in that television sensation the "reality show," bonobos engaged in the daily histrionics of their real lives are making it up as they go along. They do, however, have a script for their continuing saga, one with melodramatics worthy of a Puccini libretto and political maneuvers reminiscent of a Machiavellian treatise. The script is written after the fact, by their watchful keepers and recorded by Zoo registrar Karin Schwartz.

Photo by Rick Brodzeller

Zuri looks up at Linda.

Schwartz, who earned her master's degree in animal behavior at the University of Missouri-St. Louis, became Zoo registrar in 1990 after seven years in Zoo Pride, the volunteer auxiliary of the Zoological Society. In the early 1990s, records consisted mainly of information about acquisitions and transfers, documenting who owned what animals and where they had been sent for breeding purposes. A computer program designed to keep zoological records was only 5 years old. There were no clearly defined standards for what information should be included. Working with Milwaukee County zookeepers and Jan Rafert, curator for primates and small mammals, Schwartz developed a system for determining what was relevant.

Daily reports from keepers still reflect those protocols, with notes on animal management, training, development, sexual behavior, aggression, medical conditions and procedures, and group behavior. From that vast daily database, the registrar selects the information vital for permanent records. This information is put into a Specimen Record for each animal, which in turn becomes part of the International Species Information System (ISIS), a global sharing of computer-based information from more than 650 zoological institutions. Schwartz and zookeepers in Milwaukee had to decide for themselves what to include in that vital database. In 1990, many zoos had registrars, but there was no training program for the post. Schwartz started one. She still teaches and acts as co-administrator for the Institutional Records Keeping Course offered by the Association of Zoos and Aquariums.

"I was lucky," Schwartz recalled. "I got in on the bottom floor of a new field."

BONOBOS AS SOAP OPERA STARS

Schwartz readily admits that the data she compiles has entertainment value as well as scientific importance. The bonobos, along with the Humboldt penguins, she says, are the soap opera stars of the computerized documents known as

Specimen Records. This isn't just any old database. The records are as intricate as genealogical charts, as detailed as medical histories, as revealing as psychological profiles and as intimate as personal diaries. Even the most prosaic parts of the Specimen Records – documenting the birthplace of each bonobo and noting the locations where each has lived in captivity – are rich in content. Each wild-caught primate, essential to the genetic vitality of the captive-breeding community, also is a stark reminder of the bad old days before so many countries had signed the Convention on International Trade of Endangered Species. At the time when the senior members of Milwaukee's bonobo community were born, global trade in exotic animals was almost completely unregulated.

The Specimen Records also speak of relationship. They reveal the growing capacity of human caregivers to recognize what is happening among their charges. At first, keepers record mainly the health and breeding behaviors of the bonobos. Gradually, the story broadens and deepens. A training program designed to engage the minds of the primates is documented along with the bonobo response to this innovation. The log begins to detail the personality traits of individuals and the complexities of group dynamics. Before our eyes we see a community unfolding – honoring or challenging leaders, rearing young, enforcing standards, learning, playing, quarreling and resolving conflicts.

Zounds, What Melodrama!

We see Maringa, for almost two decades the undisputed empress of the group, in full diva mode, throwing tantrums and slyly peeking to note their effect. We glimpse the intricate etiquette that orders relationships among group members.

Photo by Mark Scheuber
Maringa and Zomi

We witness the matriarch and her "sisters" in power administering swift and painful rebukes in response to infractions of the colony's behavior code. We view Maringa's own slide from power. Here we observe the philosopher-king Lody, long the dominant male, ruling by empathy and wisdom, wielding power but always within the context of a matriarchal society. We watch younger males challenge his pre-eminence as he acknowledges his gradual physical decline. We experience tender moments between mothers and children, mischievous testing of the patience of tribal elders and compassionate interventions on behalf of weaker bonobos.

Like a social kaleidoscope, the reports make visible the ever-changing groupings that reflect temporary alliances, ambitious power plays, vicious fights and affectionate reconciliations. The Specimen Records offer more than daily evidence of bonobo intelligence. Schwartz's records also document 20 years of informed observation on beings described, along with chimpanzees, as one of the two species on Earth most like our own.

Dramatic Episodes:

As recorded by Zoo registrar Karin Schwartz, here are some glimpses of daily bonobo life as it is reported by the people who know these primates best: their keepers. Although bonobo life is intermittently punctuated by violence, sexual encounters and displays of affection are the backdrop against which the more dramatic events play out. Tempers flare, but they also quickly settle down. Buddies exchange bites and punches and are reunited. Wounds are sustained, but they heal quickly, usually without medical intervention. All in all, according to keeper Barbara Bell, the social arena for bonobos is less like a battlefield and more like a school playground with sudden squabbles that are quickly forgotten.

The Patient

Integrating Brian, an emotionally troubled male, into the group was a challenge even after keepers began acting on the advice of consulting psychiatrist Dr. Harry Prosen. When Brian went on exhibit for the first time February 19, 1998, keepers limited the experience to just a little more than two hours. They were pleased to see that he engaged in a little bit of play with other members of the group, but noted that he mostly stayed close to Lody, the dominant male who had treated him kindly. By January 7, 1999, Brian felt confident enough to bicker with Murph. Evidently, he had forgotten that Murph had powerful female relatives – his mother, Laura, and his grandmother, Linda. The ladies reminded Brian swiftly and surely that nobody (except them) picks on

Photo by Mark Scheuber

Brian

their favorite son with impunity. For a few days Brian had good reason to understand why high-ranking bonobo females sometimes are called "toe-biters." Brian keeps forgetting. At one point, keepers suggested that after spending several days with the same group, he became so confident that he became aggressive. Nevertheless, by August 20, 2001, he was correctly reading the social dynamics of a group overseen by two heavy-duty females, Maringa and Laura. Just a few days later, Ana Neema trusted him enough to allow him to stroke with one gentle finger her 10-day-old son, Bila Isia.

January 20, 2002 – Brian was upstairs with a group of 10 individuals, including three adult females. There was a large group of people in the building, which didn't seem to bother him. He has come a long way. There was no induced vomiting or self-harming behavior seen in a situation that just six months ago would have caused him lots of stress.

March 28, 2002 – Brian held his own in a volatile social grouping. He has a new attitude and a new rank (Lody, Lomako, then Brian, then Viaje).

April 22, 2002 – Brian spent the day with Maringa, Ana Neema, Linda and Kosana. The females put him in his place and roughed him up, but Brian handled it well. He used exactly the right social tactics and correct reconciliation behaviors. In other words, he spent the day grooming the females, one after another. He has adapted very well.

June 2, 2002 — Brian was placed in a group of 13 bonobos, including five adult females. He seems ready for this grouping. He took charge, organized the group and demonstrated brilliant leadership skills. Lody just sat back and let Brian run the

May 4, 2003 – While groups were being mixed and matched, Laura ripped into Brian and chewed up his hand. His pride was crushed, and he begged to go sit with Kitty. Linda chewed up Murph and Makanza. Maringa screamed non-stop at Lomako.

February 13, 2005 – Brian at 16 years old is finally acting about 11 or 12 years old. He is teasing Maringa, who he knows is slow and handicapped. He's challenging Lomako for his position. He attacked Lomako and ripped his hands, buttocks, feet and head. Brian has no mother to rein him in socially. He was put in with older females Linda and Laura. He looked miserable and was quite stressed.

The Stranger

Viaje, a new arrival who had been alone for several years in a Mexican zoo, offered a problem quite different from that posed by Brian. Things went well on January 28, 2002, when he was introduced to a group consisting of Lody, Lomako, Maringa, Laura and babies Zanga and Zomi. He acted juvenile and submissive to all members of the group and was observed holding both babies at various times. The next day, Lody led Viaje upstairs to the public exhibit area with Viaje holding tightly to the hands of the dominant male. In larger groupings, however, Viaje didn't do so well, and Ana Neema rarely passed on an opportunity to bully the newcomer. Whenever possible, Viaje avoided her.

March 28, 2002 – Viaje is uptight and depressed in a large social setting. He is becoming increasingly withdrawn. He was moved into a small grouping, away from Ana Neema.

April 5, 2002 – Four copulations seen today between Maringa and Viaje. Viaje's breeding behavior was very submissive, but he did make four attempts after Maringa backed into him. Viaje won't breed when Brian or Lomako are present in the group since he is the lowest-ranking male right now. Viaje has to be alone with Maringa. Lody can also be with them. Offering whole fruit seemed to trigger breeding behavior.

November 23, 2005 – Viaje was put together with Laura for breeding. Viaje was beside himself with excitement.

December 11, 2005 – Viaje and Brian are taking turns holding the playroom ransom. Keeper Barbara Bell doesn't think Viaje really understands the implications in this act of power and control. Viaje is just happy that Brian is his best friend today. Sex, power and door-sitting is the world of Brian right now. He's attempting a comeback, controlling one entire side of the enclosures. Tamia is ovulating and is flashing her rear end in the window for Brian. He's totally unable to function and is completely distracted by Tamia.

Photo by Richard Brodzeller

Lody

Power Plays

Now that the long-dominant Lody is beginning to decline physically, younger bonobos are vying for the male leadership role. In particular, Brian is issuing a direct challenge to Lody, but Murph and sometimes even Lomako are involved in the ongoing skirmishes for No. 1 status. Sometimes, this competition turns violent, as it did in November 2003 when Brian attacked Lody. Often the maneuvering is a matter of sexual politics.

April 29, 2002 – Brian seems to have taken hold of the No. 2 male position, right behind Lody. He struts, puffs up his hair and has a swagger to his walk. Brian took on four females today, held his ground and copulated with them all. Lomako doesn't seem to mind. He spent his day playing tag with the babies.

Photo by Mike Nepper

Brian

July 25, 2002 – Lomako tried to grab Lody while Lody was copulating with Kosana. Lody got very angry, picked up Lomako by a leg and an arm and threw him hard against the glass. Lody then repeated this action two more times. Lomako was shaken up, but seemed unharmed.

July 29, 2002 – Brian has been grabbing Kosana at shifting times and pulling her into a secluded spot for a bit of private breeding time. He did this three times today. Lody noticed this activity and, at supper time, he doubled back before Brian could and copulated with Kosana.

November 29, 2003 – Brian seems in charge of the group in the absence of the injured Lody. Brian is strutting his stuff and is very pumped up mentally. Lomako is just happy to follow along. He never had much leadership skill.

December 5, 2003 – It seems that Brian is trying to take over. When he was put in with Lody, he puffed up and went nuts. Maringa was confused. Ana Neema was panicky. Kosana was unsure. Lody was very angry. Brian was pulled out. Lody cooled down. Lomako ignored it all and ate his breakfast.

Photo by Mike Nepper

Murph

January 10, 2004 – Murph and Lomako went at each other through the mesh. Some blood drops were found here and there. Both were ready to fight.

January 18, 2004 – Murph and Brian are jockeying for alpha spot with the females. Today, Murph won due to (his mother) Laura's influence. Brian wins when Lomako is present. Lomako gladly takes the deputy role and backs up Brian. The females also are changing the guard frequently. The underdogs – Viaje and Kosana – have no clue as to the social dynamics. Lody refuses to train, refuses to move, and is keeping watch on the politics constantly. He's content to stay away for now.

February 12, 2004 – Brian stopped in front of Murph while shifting upstairs. Brian grabbed Laura (Murph's mother) and copulated with her, all for Murph's benefit. Murph got angry while Brian was grinning and laughing during the copulation.

February 15, 2004 – Linda and Murph ganged up on Brian on exhibit and ripped into him.

February 19, 2004 – Brian crawled on his belly toward Lody, begging to be admitted back into the group.

February 20, 2004 – Brian was quite humble, respectful and cautious to all. He has resumed the second rank, maybe lower status.

March 15, 2004 – Brian was run out of the all-male group this afternoon by Lody, Lomako, and Makanza. Lomako seemed very happy with this development. Brian has sunk back to 3rd or 4th ranking male. He opted to sit with Viaje, who has zero rank. Brian's pride was hurt and his ear torn.

February 28, 2005 – Brian went after Viaje this morning and bit him on the toe. Brian is a loose cannon, very unpredictable. Lomako was very upset. He was totally spooked when Brian was chasing Viaje. Lomako sat under the nest platform with a sheet on his head, shaking.

March 2, 2005 – Lomako's physical wounds (from a February attack by Brian) look better, but his mental wounds are still pronounced. He's acting withdrawn and is terrified of Brian. Brian, on the other hand, looks fine. He's begging to get back in the action.

March 8, 2005 – Brian reminds Lomako several times daily that Lomako is lower than dirt in rank. Lomako is terrified of Brian.

April 4, 2005 – Lomako and Brian were nose to nose today with mesh in between them. There were violent threats out of Brian. Lomako ignored him.

July 11, 2005 – Lody is not the powerhouse male anymore. He is not interested in group management, wants to be left alone and refuses to deal with the obstreperous Murph. He is also being pushed around now by Tamia and Zanga, his own daughter. This change in status has been coming for over a year. At closing time, Murph, Zanga and Zuri were picking on, pestering and slamming into Lody. He was confused, had no place to go and begged to go back outdoors. There seems to be a takeover attempt brewing. Lody will be kept safe for his mental well-being.

September 5, 2005 – Brian has made it firmly clear that Viaje has no business breeding with Laura. Brian was removed from the grouping. Viaje and Laura were observed copulating three or four times.

Photo by Mark Scheuber

Makanza

September 19, 2005 – Makanza started relentless teasing toward Murph. Murph gave him a few of Lody's characteristic warnings – evil stares and swats. When Makanza didn't get the message and bit Murph in the hand, Murph bit Makanza in the right foot, roughed him up and slammed him hard against the wall. This beating probably would have been administered last year by Lody if he had been feeling better.

October 10, 2005 – Makanza is annoying everyone. Nobody wants him in their group. Brian is keeping him in line with force and mental games.

October 22, 2005 – Lody should be kept quiet and calm. He can't fight anymore as he's not up for the physical challenges. He refused to go on exhibit tonight. His family sat by his side.

October 25, 2005 – Lody held back at suppertime and wanted a private spot to himself for eating. Lomako has been his faithful companion daily.

Formidable Females

While the males strut and scuffle, the dominant females remind them frequently that bonobo society is matriarchal. Ruling ladies take what they want (like extra bedding material), mate as they wish and hand out punishment as they decide, sometimes months after perceived offenses.

February 28, 2001 –Lomako was put in with Linda at suppertime. Lomako started teasing Linda right away. Linda passed off Zuri (her great-grandson, being fostered by her) to Laura and then went after Lomako. Lomako backed off and sat behind Lody, finding a hunk of coconut to inspect. No eye contact was made between Lomako and Linda the rest of the time. Lomako was still inspecting the coconut one hour later.

Linda

March 29, 2001 – Brian exhibited another aggressive outburst in morning toward a keeper. He tried to grab her ankle. He was disciplined by the keeper and by Maringa. It appears that if Brian spends several days in the same grouping, he becomes more confident and attempts aggression.

May 9, 2001 – Ana Neema has been making her bedding two to three times thicker than anyone else. This evening she took bedding off three other platforms to make quite a comfortable bed.

March 8, 2004 — Linda and Brian are together. She has him watching his back, protecting his ears and quite subdued mentally, perfect manners from Brian.

August 6, 2004 – Linda went ballistic toward Brian. She inflicted bite wounds on his hands, feet and buttocks. He was shaking, pale, and wild-eyed.

January 26, 2005 – Ana Neema and Brian are in close partnership and together ganged up on Lomako and ran him out of the group. He sustained numerous bite wounds. Ana Neema has a swollen left brow ridge and eyelid as she probably took a punch from Lomako. Brian and Ana Neema were separated, and Brian threw a huge tantrum. Ana Neema then bred Viaje six times within about six minutes. Viaje was terrified, but he performed.

January 27, 2005 – Brian was put back in with Lody, Maringa and Lomako the day after ganging up on Lomako with Ana Neema. Brian was highly uncomfortable socially.

Murph rests his chin on his hand.

March 1, 2005 – Linda held down Brian, and Murph ripped into him. Brian is not welcome in the Laura/Linda grouping. Brian was whimpering and begging for help. Brian refused antibiotics for his wounds and refused to cooperate for treatment. He was seen sitting with Kitty and rocking.

June 12, 2005 – Lody was attacked by Tamia, Linda, and Ana Neema. He sustained moderate bite wounds.

July 9, 2005 – Murph attempted to take on all the females without his mother Laura's backup. He won't be doing that again anytime soon. He came away with fingers and toes a bit ripped up and damaged pride. Later, there was an attempt to put Brian and Murph together, but both refused to cooperate.

Bonobos: Encounters in Empathy

September 2, 2005 – Lody and Brian got into an altercation. Lody was bitten on the toes, and became totally submissive in the group. Brian was quite cocky until Maringa screamed at him.

October 28, 2005 – Brian was very rough with Maringa and baby Faith. He did a few full-force tub shoves on Maringa and added a couple of kicks for good measure. His goal was to enrage Lomako and Lody, in which he was successful. Maringa was very upset, but she had no backup. (However, Brian's behavior did not go unnoticed by the new feminine forces.)

November 18, 2005 – Brian was beaten up by Tamia and Linda. Murph took them on next. This is all part of the alpha male challenge. Brian sustained bite wounds to the hands, feet, buttocks and back. Mentally, he's a wreck. He refused antibiotics and cleaning of wounds. He was feeling miserable. Kitty helped clean him up.

November 19, 2005 – Murph is acting like he's the top dog. He's acting all puffy and macho. Of course, his mother, Laura, is helping him along. Brian's wounds look okay. He's very depressed. He hung around with Kitty all day.

February 13, 2006 – Lomako was observed copulating with Laura. Laura's son Murph had a total meltdown spell with tantrums and loose stool.

The Baby Book

Central to bonobo communal life is the vital role of offspring. Bonobo babies born at the Milwaukee County Zoo immediately cling to their mothers. They are carried constantly. Infants fostered here are greeted with enthusiasm and cuddled by males and females. The Specimen Reports bear witness to the healing influence of infants on Brian, the most disturbed member of the group, and to the gentle nurture provided to youngsters by Lody, the alpha male. The Baby Book also reveals the way bonobo infants move toward independence from their mothers, some easily and others surrounded by maternal histrionics.

Photo by Mark Scheuber

Brian plays gently with Zanga Mokila

March 3, 2000 – Lody was observed taking 14-month-old Zanga from her mother, Laura. He played with Zanga for over an hour, tickling, laughing, and cuddling together. Laura seemed fine with the arrangement. Lody then delivered Zanga back to Laura.

March 16, 2000 – Laura allowed Brian to hold baby Zanga. Brian really seemed to enjoy it.

May 25, 2000 – Laura, even though she had an infant of her own, did everything possible to welcome Zuri, a keeper-reared infant rejected by his mother at his birth at the San Diego Zoo and transferred to Milwaukee for socialization. She moved her nest close to Zuri, gestured to him to follow her, etc. Zuri kept his distance at first, but then did have a few positive interactions. Some playtime was seen, also.

May 27, 2000 – Lody, the alpha male, quickly scooped Zuri up for a few hugs. Lody then calmly hung out with Zuri, offering comfort but not forcing himself on Zuri.

June 1, 2000 – Brian, who did not have a normal bonobo infancy himself, gently held Zuri all day long. Keepers considered the behavior very positive for both adult and infant.

June 10, 2000 – Ana Neema, new to the group herself, is spellbound with Zuri. She cuddles, rocks and attempts to nurse him. Zuri is enjoying the contact.

June 11, 2000 – Brian is exhibiting new play behavior. He plays non-stop with Zuri and Ana Neema. Zuri is a huge positive in Brian's life.

June 14, 2000 – After an earlier training session designed to strengthen the bond between adult female and infant, Ana Neema continues to show maternal behavior about 75% of the time toward Zuri when they're together. Zuri shows "follow mom" behavior 50%-75% of the time toward Ana Neema.

June 29, 2000 – Brian is very jealous over Zuri now. Brian pouts, sulks and throws things when a keeper or another bonobo shows attention to Zuri.

July 7, 2000 – Brian was very aggressive toward Zuri. Brian kicked Zuri and yanked him off the mesh. Brian was separated from Zuri, and Zuri was put in with Maringa, who treated him very well.

July 23, 2000 – Linda is in control of the care of Zuri. She dictates who can/cannot hold Zuri. Zuri is thriving under her attention.

August 23, 2000 – Zuri went up to the exhibit for the first time, traveling on Linda's back. Zuri was confident, playful and enjoyed the new surrounding. His skills have improved drastically. He's not the frightened little blanket-clutching baby anymore.

October 20, 2000 – Linda carried Maringa's baby Zomi quite a bit today. Linda also had her own baby, Zuri, hanging on as well. Maringa seemed fine with this arrangement. Linda is turning out to be a brilliant caretaker.

October 29, 2000 – Makanza (a 6-year-old fostered since his mother died) seems to have traded "mothers." Due to the behavior of Murph, who at age 10 has entered the "brat stage" of bonobo development, Makanza now prefers Maringa over Murph's biological mother, Laura, who has acted as a foster mother to Makanza for the past five years. Maringa now provides Makanza with comfort and nurturing. Laura seems to have accepted this transfer of affections.

April 19, 2001 – Lody played with Zuri, Zomi (Maringa's 21-month-old daughter) and Zanga for approximately one hour.

July 16, 2003 – Zanga (4 ½ years old) split off from her mother, Laura, and went to play with Zuri upstairs all on her own. Laura hardly noticed but kept checking the door. Lody babysat Zanga. Three hours went by, and Zanga then came back downstairs on her own and was reunited with Laura. Zuri (5 years old) was split off

Photo by Richard Brodzeller

Zuri stays close to an animated Linda.

from Linda. A few whimpers were heard from Zuri. Brian looked after Zuri. The experience was positive for all.

October 15, 2003 – Zuri is spending the night without Linda. He shifted away from her, and it didn't seem to bother her.

December 8, 2003 – Lomako was observed holding and playing calmly and gently with baby Deidre.

February 1, 2004 – Zomi split off from Maringa for about 45 minutes. Maringa failed to notice this separation for 10 to 12 minutes as she was busy trimming Lody's hair. She then went ballistic, belly-flopping, whimpering and overcome with emotion. Zomi was busy playing. She paid no attention to the maternal outburst. When her tantrum was done, Maringa returned to trimming Lody's hair.

March 22, 2004 — Zuri is spending much time with Laura. Laura is incredible with him and gives him a type of mothering different from Linda's. Zuri needs the touching and pampering that Laura provides.

April 29, 2004 – Zuri has joined the Lody and Zomi grouping. Both Zuri and Zomi have forgotten that they don't have a mother in the group. They're having a great time. Zomi's also learning that her mother isn't there for immediate backup when the play gets rough.

April 30, 2004 – Three days after her nearly 5-year-old daughter was separated from her for weaning, Maringa threw a full-blown hissy fit from 8:15 to 10 a.m. today. First there was pouting, then whimpering, then a "seizure" on her back (with one eye open to observe the effect of her emoting on keepers), then, for a finale, a flip over to her belly with rapid jerking and shaking. When all that failed to get attention, she went to breakfast.

(As Bell interprets Maringa's actions, the fits are thrown principally for the benefit of the keepers to let the diva's support staffers know that their arrangements are being met with extreme displeasure.)

March 24, 2005 – Zanga has increased her time away from her mother, Laura. She's getting increasingly independent. A small genital swelling was observed.

High Jinks

Not all bonobo life is power politics and pro-creation. Often, everybody's just out for a good time. Many happy occasions, of course, are not immediately understandable outside the bonobo group, but some activity is obviously just for fun.

March 25, 2001 – Ana Neema and Lomako stole the blood-pressure cuff and hid it on a platform with a lot of giggling and smiling. Lomako gave in and traded the cuff (with a zookeeper) for a chunk of licorice.

March 28, 2002 – Kitty (who is elderly and blind) has been very active for the last three weeks. She plays, runs around, socializes, suns herself and doesn't seem to sleep. She recently was seen bouncing a ball and laughing.

Photo by Richard Brodzeller

Kitty

July 9, 2002 – Lody was reintroduced to finger painting after a long hiatus. He used his feet while laughing non-stop and used wood wool to dab on paint.

November 18, 2002 – Bonobos caught a live squirrel today and played with him for hours. Lody and Brian refused to let it go, trading it back and forth. Lody held the squirrel up to his face and made a "happy play face" again and again. Brian smacked it around. The squirrel was retrieved by a keeper and sent to the hospital. After its condition proved to be healthy, the squirrel was released near Lake Evinrude.

Photo by Richard Brodzeller

Barbara Bell provides materials so Lody can fingerpaint for recreation.

November 23, 2005 – Lomako rolled, danced and jumped in the new snow. There were lots of snowball fights this morning.

January 21, 2006 – Zuri, Makanza and Brian were given time outside. They all had fun playing in the fresh snow.

February 13, 2006 – Brian stole the hose for the first time. He got caught by Keeper Claire Richard and dropped it quickly.

Tender Senior Moments

Only the bonobo babies have been observed receiving more tender loving care than Kitty, the undisputed senior member of the group at the Milwaukee County Zoo. Blind, deaf and subject to epileptic seizures, Kitty has benefited at one time or another from the gentle assistance of most of her fellow bonobos. In the case of Brian, she has provided much needed soothing and calm.

Photo by Mark Scheuber

Kitty

September 28, 1997 – Kitty introduced to Brian. All went well. Linda was already with him at the time of introduction. There were gentle interactions among the three.

July 12, 1998 – Brian helped Kitty outdoors this morning. He was very patient and gentle in guiding her along. It's a milestone for Brian to be the helper instead of the one needing assistance.

May 31, 1999 – Lody carried Kitty to the outdoor ladder this morning when she was obviously very disoriented and panicky.

November 26, 1999 – Both Brian and Kitty sat in front of the seasonal affective disorder light today. Kitty really liked it.

March 15, 2000 – Gentle, caring, compassionate behaviors were seen out of Lomako toward Kitty. He helped her indoors today. His patience level has increased dramatically.

March 4, 2001 – Even the *enfant terrible* Murph is moved to kindness by Kitty. After she had a seizure, he went over to nudge her gently.

July 24, 2004 – Kitty had a seizure this morning at 9:25 a.m. Makanza chose to sit with her today and had his arms around her most of the morning. This level of empathy, compassion and mental maturity shown by Makanza at age 10 is a byproduct of the diversity of the group.

April 18, 2005 – Finally, keepers were successful in getting Kitty outside. She had been staying in her pen since March 27, when she was bitten by Ana Neema. Kitty was happy to be outside and was hugging everyone.

Medical Milestones

Few results of the bonobo training program at the Milwaukee County Zoo have been as beneficial as those associated with the health of these primates. The Specimen Reports are full of notations regarding medical procedures. Particularly important has been information gathered in Milwaukee about fetal development, respiratory infections and heart-related problems. The training program is credited by Zoo veterinarians with a transformation in the way health can be monitored and care can be given.

Photo by Richard Brodzeller

Bell prepares a bonobo for an ultrasound exam. The bonobo lies in a protective area above Bell.

July 22, 2002 – Laura had an obstetrical ultrasound done. All looks very normal. Baby is possibly a female. The images were the best and clearest images that have been taken so far. Perfect images of the infant's face were seen, with eyes clearly blinking, infant swallowing, etc.

August 14, 2002 – Laura allowed a 3D ultrasound to be performed and had perfect manners. She tired out after about 50 minutes, but was incredibly cooperative.

August 23, 2002 – Laura gave birth to a daughter, Claudine. Baby was born at approximately 11:40 a.m. By the afternoon, the baby was clean and nursing, but the umbilical cord was still attached. About two feet of cord is dragging on the ground.

November 26, 2002 – Echocardiograms were done on Lody and Lomako and Ana Neema and Maringa. Everyone cooperated completely. When they heard that "Leann's coming" (Leann Roth Beehler, the ultrasound technician), they split apart, got ready to scan and performed brilliantly. Lody almost fell asleep during his scan. The procedure was done stress-free.

August 5, 2005 – Brian did a perfect layout in the chute and actually worked well on positioning for medical procedures. He outdid himself and did a perfect hold on the blood-pressure sleeve. He was rewarded heavily (with food treats).

Ripples in the Rain Forest

Photo by Gay E. Reinartz

Wild bonobo at Lola ya Bonobo, a bonobo orphanage in Kinshasa.

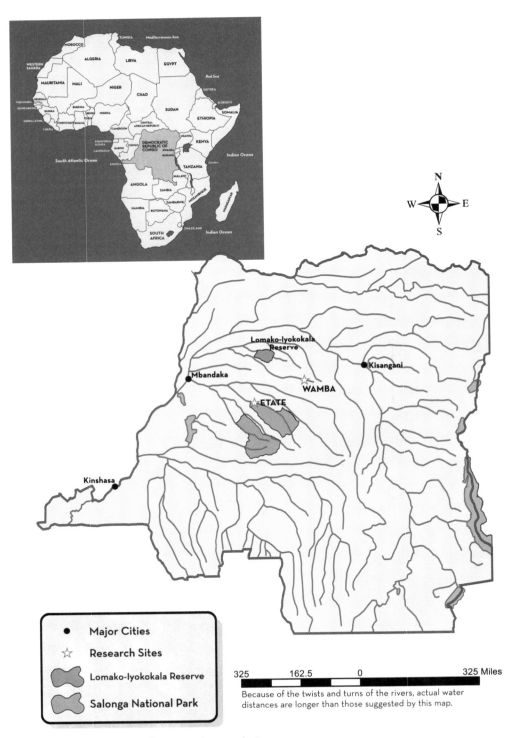

Map of Democratic Republic of Congo with research sites

Bonobos: Encounters in Empathy

CHAPTER 7

Ripples in the Rain Forest

Photo provided by Gay E. Reinartz

Dr. Gay Edwards Reinartz takes a dugout canoe called a pirogue into Salonga National Park, accessible mainly by water.

The humid air above the Salonga River shimmers with rainbows and hums with malarial mosquitoes. In a motor-powered dugout canoe locally known as a pirogue, Dr. Gay E. Reinartz, conservation coordinator for the Zoological Society of Milwaukee, is on her way to work – a commute up river that takes almost two days when everything goes smoothly, four when it does not. It is not a dull commute. There almost certainly will be a tropical downpour.

The Democratic Republic of Congo (DRC) is a land of heart-lifting beauty and soul-wrenching poverty. It is one of the most magnificent places in the world to work, Dr. Reinartz says, and the most difficult. At its heart lies Salonga National Park, a vast equatorial rain-forest preserve half the size of Wisconsin that forms the heart of the bonobos' range. Because roads were among the casualties of the country's economic collapse and two civil wars, travel to the park is mostly by air or water. Recently, an airfield has been built in the middle of Salonga. However, the landing strip is a four-day walk from Etate, the Zoological Society's research station near the junction of the Salonga and Yenge (YEN-gay) Rivers. At this station the Zoological Society is conducting ecological research to help discover whether the park harbors a self-sustaining bonobo population and what habitats these primates prefer.

Mounting an expedition at least twice a year, once in spring and once in fall, is a major project in itself. The Zoological Society must raise about $20,000 for each research mission. Much of Dr. Reinartz's time each year in Milwaukee is spent completing grant applications, writing reports for funding agencies, making PowerPoint presentations to supporting organizations and meeting with potential donors. Then she embarks. The trip begins with a 20-hour series of flights from Milwaukee to the DRC capital, Kinshasa. She spends about a week touching base with government officials and members of the academic and conservation communities. That includes consulting with representatives of the Institut Congolais pour le Conservation de la Nature (ICCN, i.e., the Congolese Institute for the Conservation of Nature under the auspices of the Ministry of the Environment,

Water and Forest). The Zoological Society is a principal partner with ICCN, the agency overseeing national parks and reserves. The primary responsibility of the Zoological Society in Congo is to support and enhance the technical capacity of the ICCN, which requires finances and training from international conservation groups to manage protected areas.

GUNS ALONG THE RIVER

It is a two-hour flight from Kinshasa, where the Zoological Society has an office, to Mbandaka, the city closest to the patrol post and research station at Etate. In Mbandaka, Dr. Reinartz organizes the mission with the small staff of Congolese nationals who form the expedition team that travels with her to Salonga. Loketshi Lokuli Mira (loh-KET-she loh-KOO-lee MEE-rah) is lead staffer in charge of organizing the long pirogue journey and the camp at Etate. Together Mira and Dr. Reinartz locate and purchase enough supplies, including the barrels in which to transport them, to last two months in the forest. Over the course of the year, the expedition team delivers to park headquarters at Watsi Kengo (WAHT-see KEN-goh) more than 1,300 gallons of fuel to power the outboard motor (donated by the Zoological Society) used on anti-poaching patrols of the park's waterways.

Sometimes, adventures begin even before the pirogue leaves the dock at Mbandaka. On one occasion, the expedition team was sharing a lunch of roast chicken while waiting for the last member of the crew to arrive. Five heavily armed naval police boarded the boat and demanded a few barrels of fuel. Dr. Reinartz told the would-be hijackers that she was working for the government and needed every bit of the fuel to complete a long journey. She added, however, that although she had no gas to spare, she did have more chicken than she needed. Would they like some lunch? She offered the leader her portion of the meal, which custom would not allow him to refuse. The men ate, bummed some cigarettes, and left.

Photo provided by Gay E. Reinartz

Zoological Society staff unload barrels of fuel and supplies at a patrol post in Salonga National Park.

To reach the remote outpost of Etate, river pilots Moyembi Vincent (moh-YEM-bee) and Monkoli Redo (mohn-KOH-lee RAY-doh) guide the motorized dugout day and night for more than 375 miles up the Congo, the Ruki (ROO-kee), the Busira (boo-SEE-rah) and – finally – the Salonga Rivers. Nevertheless, stepping into that open boat for a voyage of at least 48 hours is a much anticipated moment for Dr. Reinartz. That is true even though the only rest stops en route are jungle bushes and much of the trip is likely to take place in torrential rain. Risks are carefully calculated before each mission. The Zoological Society team is accompanied by two military escorts who are there to intervene in case the expedition encounters poachers, often themselves rogue members of the military. This is, after all, a

Bonobos: Encounters in Empathy

country where sporadic fighting continues to take its toll in eastern areas and where international peacekeepers were invited to ensure order during the elections of summer 2006. However, Dr. Reinartz insists that the destination is worth almost any amount of discomfort. For her and for the Zoological Society of Milwaukee, Etate represents the culmination of more than two decades of work. Were it not for them and the ICCN park guards, the outpost might still be a camp for poachers, not the center of what is now an international effort to protect bonobos in the heart of their natural home.

Like almost every other element in this story of Milwaukee's relationship with bonobos, this journey and its destination are the consequences of informed caring. Dr. Reinartz arrived in the Milwaukee area in 1980 after earning her master's degree in conservation genetics at Duke University. In 1983, she began work for the Zoological Society of Milwaukee as a consultant, providing genetic analysis of the Zoo's Dall sheep, snow leopards and Siberian tigers, all success stories in Milwaukee's captive-breeding program. Preserving genetic diversity in finite populations had been the topic that drew Dr. Reinartz to graduate school; so she was immediately attracted by the opportunity to analyze the pedigrees of animals bred in United States zoos and to determine how limited gene pools affected their survival.

AN AMAZING OPPORTUNITY

By October of that year she was acting as the Zoo's registrar, charting the histories and relationships of many of the Zoo's specimens. Then she noticed, in a zoological publication, an ad from Dierenpark, a small private zoo at Wassenaar outside Amsterdam in the Netherlands. The owners were seeking a single home for their bonobos and gorillas. Dr. Gil Boese, then Zoo director, was immediately interested because the Zoo was looking for gorillas. The bonobos seemed to be a fortunate bonus. At the time, there were very few bonobos in North America. In 1986, the San Diego Zoo was home to North America's largest captive colony, 12 animals. There were 10 in research facilities in Georgia. Four other wild-caught bonobos lived at a zoo in Morelia, Mexico.

Dr. Reinartz performed the first pedigree analysis of the world's captive bonobo population and discovered that most bonobos in the United States were related, born as the result of breeding a precious five wild-caught animals, one of which had died in 1980. Milwaukee had a chance to obtain seven of the rare primates – none of whom were related to the U.S. population. Naomi, a 4-year-old female, and Lomako, a 2-year-old male, both were born at Wassenaar, but the remaining five had been captured in the wild. If Dr. Reinartz and Dr. Boese could negotiate the purchase and add these bonobos to the existing gene pool in the United States, they could increase dramatically the genetic diversity

Photo provided by the Wassenaar Zoo and Sam La Malfa

Lomako on the left and Naomi on the right were born at Wassenaar.

available for captive breeding. There was a fierce bidding war between the Milwaukee County Zoo and buyers in Saudi Arabia who had virtually unlimited funds. Although Dr. Boese's final offer was lower than that of the opposing bidders, he clinched the deal with an argument that recognized the loving care that had been given the bonobos at Wassenaar. Zoo owners Jan and Hanneke Louwman had nurtured the infant bonobos with the kind of close attention and constant cuddling that their natural mothers would have provided.

Dr. Boese told the Louwmans: "We will guarantee that we will tell 1.5 million Zoo visitors a year that these animals came from the Wassenaar Zoo…" The notion that bonobos in Milwaukee would keep alive the story of Wassenaar was the deal maker. The Dutch negotiators said "yes."

The next possible hurdle was winning the necessary import permits under the Convention on International Trade in Endangered Species (CITES) and the Endangered Species Act. Ordinarily, the sale of the bonobos at Wassenaar would have been prohibited because five of them had been caught in the wild. The Zoo met objections to the proposed importation by pledging to begin a captive-breeding consortium in the United States and by arguing successfully that the five potential founder bonobos were needed to enhance the very limited gene pool in this country.

On November 20, 1986, the Zoo received a permit to import the animals. By December 3, the seven primates were getting used to their new home at the Milwaukee County Zoo.

To Survive and Thrive

There were exciting possibilities ahead. By this time Dr. Reinartz was associate zoo director for science and conservation. Over the next two years she worked with the American Zoo and Aquarium Association (now the Association of Zoos and Aquariums, AZA) to develop a Species Survival Plan (SSP) for bonobos in North America. She visited zoos and research facilities and consulted with curators and keepers across North America to shape the plan approved in 1988 to develop a self-sustaining captive population with maximum genetic diversity. Dr. Reinartz was appointed species coordinator. By 1998 the SSP Committee in this hemisphere had forged an agreement with its European counterpart, the European Endangered Species Programme (EEP), to produce the first global master plan for

Photo provided by Gay E. Reinartz

a species of great ape. Since that time, all captive bonobos outside Africa have been managed as a single gene pool. That cooperative approach has allowed preservation of an astounding 96% of the genetic diversity contributed by wild-caught founder animals, Dr. Reinartz estimates. By 2006, the joint management effort had led to a North American population of 84 bonobos in 10 zoos and a European population of another 76 bonobos in nine zoos. Another 47 bonobos now live at Lola ya Bonobo (Paradise for Bonobos), a

Lola ya Bonobo is an orphanage in Kinshasa, capital of the Democratic Republic of Congo.

forest sanctuary near Kinshasa. With 21 bonobos, Milwaukee is now home to the largest captive-breeding group in the zoo world. The Columbus (OH) Zoo and Aquarium and the Wilhelma Zoo and Botanical Garden in Stuttgart, Germany – with 16 bonobos each – also host large breeding colonies.

While this work was going on, Dr. Reinartz – who was engaging in her own personal crash course on the species – discovered how relatively little had been published on the subject and how much remained to be learned. For one thing, no one had any real idea of how many bonobos actually remained in the wild. Their natural home was remote. Zaire's history was punctuated by political violence. Compared to scholars studying the more numerous gorillas and chimpanzees, relatively few researchers followed bonobos into the rain forest. A field study of the species was being conducted at Wamba (WAHM-bah) in the northeastern part of the rain forest. However, Japanese researchers led by Takayoshi Kano of Kyoto University had been working there only since 1974. An American team led by Randall Susmann of the State University of New York at Stony Brook was working in the Lomako Forest a few hundred miles away. In the popular media, the bonobo was virtually unknown. Frans de Waal's popular book *Bonobo: the Forgotten Ape* was not published until 1997.

For Dr. Reinartz, field work in Zaire was not an option. She had a young family. Both she and the Zoological Society had limited financial resources. She began seeking graduate students or post-doctoral students already in the field. By offering supplemental grants for their work, she hoped to gain insights that would guide future conservation work as well as help primate keepers provide the best possible living conditions for captive bonobos. Providing supplemental grants, however, proved to be a fragmented approach to conservation. During the 1990s, a number of researchers had to abandon the field during political unrest. As they finished their advanced degrees, student researchers moved from the field into academia, usually without the means of supporting additional conservation work.

In 1995, the Zoological Society of Milwaukee published an action plan for preserving bonobos in the wild.

The Only Action Plan

Dr. Reinartz, however, was in communication with excellent sources of reliable information: long-time researchers who studied bonobos in Zaire. Almost everyone who was engaged in such study was a member of what was then called the Bonobo Task Force of the World Conservation Union (IUCN). At a series of international conferences, primatologists Dr. Nancy Thompson-Handler and Dr. Richard Malenky, co-chairs of the Bonobo Subgroup of the Primate Specialist Group of IUCN, were asked to compile recommendations for conservation based on what was then known about the species by scientists from around the world. The Zoological Society of Milwaukee became a clearinghouse for contributions to the project. Dr. Reinartz found herself co-editing the resulting document, which was published by the Zoological Society in 1995. *Action Plan for Pan paniscus: Report on Free Ranging Populations and Proposals for their Preservation* became the first internationally recognized plan for bonobo conservation.

The action plan made painfully clear that in the one area of Zaire set aside as a refuge for this endangered species – Salonga National Park – it was still unknown whether there were any bonobos. Although Salonga was Africa's largest tropical rain-forest preserve and had been listed by the United Nations as a World Heritage Site for its outstanding treasure-trove of rare species, protection for both plants and animals had been more of an ideal than a reality. There was only anecdotal evidence that bonobos actually lived in the forest. In the quest for field sites, most researchers had bypassed the park altogether. Outside the park, logging interests were still cutting roads into the interior. Inside the park, elephants and other large mammals were being slaughtered for the ivory and bushmeat markets.

SURVEYING SALONGA

As Dr. Reinartz completed work for her own doctorate in conservation ecology and genetics of bonobos at the University of Wisconsin-Milwaukee, the Zoological Society decided it would support the action-plan recommendations. Dr. Reinartz recommended that the Zoological Society focus on Salonga since so little work had been done there and since it was critical to know whether there actually were bonobos living in the vast protected area. The Zoological Society formed a partnership with the Institut Congolais pour la Conservation de la Nature. By 1997, the ICCN and the Zoological Society, in collaboration with the Royal Zoological Society of Antwerp, were ready to send a team of researchers to Salonga to look for signs

Photo provided by Gay E. Reinartz.

of bonobos. The timing was right. Long-time dictator Joseph Mobutu Sese Seko had just been deposed. Laurent-Desire Kabila took over as president. Hopes for peace were high. The country was renamed Democratic Republic of Congo. Dr. Reinartz traveled to Kinshasa to assemble a research team and to coordinate collaboration with the ICCN, getting assistance from the U.S. Embassy. There were rumors of civil unrest in the eastern part of the country, but Belgian researcher Ellen Van Krunkelsven from the Antwerp Zoo and

In 1997 Inogwabini Bila Isia, Ellen Van Krunkelsven and Dirk Draulans went on a research mission looking for bonobos in Salonga National Park.

her Congolese counterpart, Inogwabini Bila Isia (ee-NO-gwah-BEE-nee BEEL-ah EESS-ee-ah), were hired by the Zoological Society to conduct an exploratory survey.

The team headed out in a pirogue up the Congo River. They had no satellite phone, not even a shortwave radio. They were traveling upriver into a remote part of the rain forest. They didn't know what they would find or what might find them. Two weeks went by. Three. Dr. Reinartz waited back in Wisconsin for some word of their whereabouts. There was none. Finally, after about four weeks, she heard from them. Good news. There were signs of bonobos at all four sites. The foray became the first scientific mission to the northern sector of Salonga that would confirm the presence of a resident population of the endangered primates. Unfortunately, the research team also brought back discouraging news. In this relatively tiny area of the park, they had found 43 hunting camps.

WAR AND RUMORS OF WAR

In August 1998, before Dr. Reinartz could launch a more systematic, park-wide survey, civil war broke out. Before a series of peace accords were signed in 2002, fighting raged across the country, killing at least 4 million people and displacing as many as 8 million, according to reports from the United Nations and international news services. Many call the conflict Africa's First World War. Yet during this time, the Zoological Society was laying plans and raising funds for a large-scale exploration of the forest. Dr. Reinartz traveled frequently to Kinshasa to strength-

Photo provided by Gay E. Reinartz.

en ties with conservationists there. Educational materials were prepared, published and distributed. With the help of the United Nations Peacekeeping Mission in Congo, Inogwabini and ICCN field staff in 2000 delivered emergency supplies to park guards and assessed the war's effects on Salonga. In that same year a research station staffed by Congolese was established at Etate on the Salonga River. Millions of people were on the move, still fleeing the civil

The Zoological Society's research station, Etate, has several thatched-roof buildings along the Salonga River.

unrest and desperate for the food, shelter and fuel the forest represented. Yet because of its remoteness, most of Salonga National Park was untouched by fighting. During this period Dr. Reinartz was spending frantic days in Milwaukee and Kinshasa. Prevented from travel to Salonga by restrictions resulting from the war, she pursued other avenues of conservation. It was during this period that she negotiated a contract with UNESCO and the United Nations Foundation to administer a grant on behalf of ICCN to pay the Salonga park guard salaries. By 2001 she was accompanying Inogwabini to the rain forest, a fact that does not go unnoticed among her Congolese colleagues.

"When I meet someone new," she said, "and they hear that I was in Salonga in 2001, you can see them stop and do a quick calculation. 'That was during the war,' they will say, and immediately I see a change in their regard."

Inogwabini and Dr. Reinartz together shaped the field program known as the Bonobo and Congo Biodiversity Initiative. He served as first field director.

"We learned from each other," she said. "There was a real brotherhood of conservationists during the war years associated with the UNESCO project. Together, we would meet in Kinshasa and keep things going as best we could. Inogwabini did one of the first assessments of Salonga – the needs of the park and of the people who were trying to protect it. That made it possible for us (at the Zoological Society) to target the most important conservation priorities. He initiated me into the field, and I initiated him into additional science."

What began as a search for bonobos has grown into a combination of ecological research and park support. Cooperation from Congolese authorities, academics and conservationists has been good, she said. Over the years, the Zoological Society has earned a great deal of respect for its work, she said. It helps that the Society found funds to pay the park guards, whose consistent dedication to their work long had outstripped their compensation. When they first received salaries from the Zoological Society, most of the guards had not been paid for years. The Society's reputation also has benefited from the goodwill created by other important accomplishments:

Photo provided by Gay E. Reinartz.

Etate guards with Dr. Gay E. Reinartz.

- The "Action Plan for *Pan paniscus*" published in 1995 by the Society is the first single-species action plan to be sanctioned by the World Conservation Union.
- The Society-sponsored exploratory mission in 1997 offered the first scientific proof that bonobos actually lived in the park. Dr. Reinartz and her colleagues were the first to go into the Salonga and to search the park systematically for signs of bonobos and to catalogue other forest characteristics vital to their habitat.
- To stop illegal boat traffic, usually accompanied by poaching, the Zoological Society equipped regular patrols on the Yenge River in November 2000. Only one week after stringing a cable barrier across the river, ICCN guards reported that they had confiscated 12 shotguns and a load of bushmeat. Under the auspices of the Congo Basin Forest Partnership, the Society is continuing to increase the anti-poaching patrols on the Yenge and Salonga Rivers.
- In December 2003, the Zoological Society sponsored the first formal paramilitary training for Salonga Park guards provided by members of the Congolese military. The six-week course was paid for solely by the Zoological Society. Fifty-four guards and six park wardens attended.

Photo provided by Gay E. Reinartz.

Weapons and snares were confiscated during anti-poaching patrols started by the Zoological Society.

- Guards hired by the Zoological Society and working out of Etate have initiated a vigilant anti-poaching program in an area where previous protection existed only on paper. A team of guards is directed by Mboyo Bolinga (em-BOY-oh boh-LIN-gah), chief of the Etate patrol. The guards are part of the Watsi Kengo (WAT-see KENG-oh) mobile anti-poaching unit that in 2005 confiscated nearly 3,500 metallic snares, 2,500 nylon snares, shotguns and spears.

- The Society has equipped Etate with uniforms, a digital camera, a satellite phone, a communication radio and solar panels and batteries, camping supplies, food, medicines, buildings and research equipment.

Bonobos: Encounters in Empathy

- The Zoological Society employs 21 supplementary guards who augment Salonga's undermanned force at patrol posts protecting bonobo populations discovered during the Society's site surveys.

In 2000 the Zoological Society research teams, first led by Inogwabini and then by Dr. Reinartz, began a systematic survey of different sites throughout the park. Etate was the first area sampled. Using a method of cutting narrow paths, called transects, into the forest, they searched for signs of bonobos, other large mammals and human hunting activity. Over the next 18 months, the Zoological Society team discovered five populations of bonobos in the northern and southern sectors of the park. Etate was one of the more densely populated sites. There

Photo provided by Gay E. Reinartz.

A bonobo footprint (in the square) found in the forest is shown with a pen for scale.

also were areas – containing either marginal habitat or intense hunting activity – which were totally devoid of bonobos. With evidence of bonobo residence in the forest, Congolese conservationists and their international colleagues can make stronger arguments to preserve the park from development in a country where extraction of natural resources such as timber remains a powerful engine of economic growth. The possibility of seeing bonobos in Salonga could attract the kind of ecotourism that generated significant income in eastern Congolese parks before the war.

A Bonobo Swinging His Feet

In 2003, the Zoological Society began to focus attention back to Etate to study bonobos there, to describe habitat characteristics in depth and to bolster anti-poaching and patrol efforts in the region. The team refurbished worn-out thatch buildings and settled in to make Etate home base for the Zoological Society in Congo. Therefore, there was more involved than a personal thrill on May 10, 2005, when Dr. Reinartz and the rest of the survey team looked up into the rain-forest canopy and saw – high above them but clearly visible – a large male bonobo sitting on a branch and swinging his feet. It was near the end of the research mission, only two days before Dr. Reinartz was to begin the journey back to Wisconsin. On nine previous missions, she had seen bonobos only three times, mere glimpses of disappearing animals.

"I've covered many miles in the park without seeing bonobos," Dr. Reinartz says. "For two years we were going on surveys constantly in new areas, new areas, new areas, and not until we were in the second year of research was there a direct sighting."

This trip was different. Researchers extended the grid of transects in the Etate sector where they already had confirmed that bonobos fed and nested. Often laboriously hacking paths in thick underbrush, they explored new territory. An amazed team recorded another three sightings:

- **April 27** – About a mile beyond the end of one of the station's established research corridors, the team ran into a large group of bonobos – perhaps 8 to

13 individuals – calling, coming down through the trees to look at the human intruders, moving through the forest all around the researchers for an incredible five to 10 minutes.

- **May 8** – In a swamp forest far away from the first site, a smaller group of bonobos dabbled in shallow water.

"We could see the sunlight hit their backs," Dr. Reinartz recalled. Creeping quietly along, led by Mboyo Bolinga, researchers followed the group for more than an hour, only to realize that the bonobos had circled around them.

- **May 10** – Once more the search entered an area not previously surveyed. Accompanied by patrol chief Mboyo Bolinga, forest guide Isomana Edmond (ee-soh-MAH-nah ed-MOND), two guards and her adult son, Nathaniel, Dr. Reinartz pursued an old elephant trail southwest of Etate.

Suddenly, Mboyo Bolinga and Isomana Edmond spotted bonobos in the canopy. For the next half-hour – until the humans had to return to the research station – the encounter continued. At first team members slowed their movements and hushed their voices.

From rustling movements in the trees and calls, researchers concluded that they were in the presence of at least three or four bonobos. They spotted a female moving along a limb. The big male settled down on his branch.

Photo provided by Gay E. Reinartz.

Bonobos (circled) appear as dark shadows in rain-forest trees and thus are hard to capture on film.

To everyone's surprise, the bonobos stayed where they were. Eventually, team members began speaking in normal voices. The humans moved about freely on the forest floor. Still the bonobos did not flee.

At the base of a tree lay the body of a dead infant bonobo. Mothers carry their dead babies for a while, Reinartz said, but the mother may have dropped this one by accident. Reinartz speculated that the bonobos might have stayed in the area because of the dead infant.

For whatever reason, the primates were close enough to photograph, a rare occurrence. Getting pictures, aiming straight up at a subject that is black against the sky, is a challenge, but this time it was possible. Mboyo Bolinga was particularly eager to take a picture.

"Obviously, they see bonobos more frequently than I do," Dr. Reinartz said of her colleagues, "but because we have a camera they are so excited they can barely speak. I'm enthusiastic because they are so excited. Their ancestral home is this area. They have a lot more invested in this conservation effort than we do. They put their lives on the line every day to protect the forest against poachers."

Bonobos: Encounters in Empathy

Nevertheless, Dr. Reinartz wondered at the jubilation over this last encounter. It was an exceptional experience for her, but why were the Etate regulars so thrilled? "You see bonobos all the time," she told Mboyo Bolinga. "Why are you so excited?"

Mboyo Bolinga has seen the photographs Dr. Reinartz takes in the copies of *Alive*, the Zoological Society's magazine, which she brings to the station on her missions. He knows that the magazine represents the Society that established the research station and keeps it going. He replied, "You're going to tell the world that there are bonobos at Etate."

She responds, "That's exactly what I'm going to do."

TELLING THE WORLD

Indeed that is what Dr. Reinartz continues to do. At conferences throughout the world – most recently the 21st Congress of the International Primatological

Photo provided by Gay E. Reinartz.

Exploring bonobo habitat, Zoological Society staff measure the diameter of a tree at breast height and find it is 15 feet.

Society in Entebbe, Uganda, in 2006 – she talks about what is happening in Salonga. It is vital, she declares, that the international community direct attention and commit resources to preserving what is now the world's second largest rain-forest park. The vast Congo River Basin, of which Salonga is a part, sprawls across more than 770,000 square miles, forming a contiguous rain forest second only to the Amazon. In a stream of interviews with the popular press and articles she has written for scientific journals, Dr. Reinartz keeps Salonga's bonobos in the public eye.

She and Dr. Boese traced the evolution of the Zoological Society's conservation program for *Primate Conservation: The Role of Zoological Parks*, a 1997 publication of the American Society of Primatologists. When the Brookfield Zoo hosted a conference on The Apes: Challenges for the 21st Century in May 2000, she arranged for Inogwabini, who then headed the Society's research station in Congo, to share the podium with her to give the keynote address. To get Inogwabini into the country, they waded through the six months of paperwork required to obtain a visa that permitted his appearance. He was the only Congolese speaker.

For Great Apes and Humans: The Ethics of Coexistence, a 2001 offering from the Smithsonian Institution's Scholarly

Photo by Richard Brodzeller

Dr. Gay E. Reinartz

Press, she joined coordinators of the other great ape species survival plans and AZA staff in making a case for the ethical role zoos play in preserving endangered species. A similar theme emerges in her 2003 chapter on bonobos for *The Great Ape Project Census: Recognition for the Uncounted.*

In that chapter, she wrote: "While recently researching the genetic diversity of bonobos, I analyzed the DNA of all the wild-born bonobos in North American zoos and came upon a startling discovery: At least half of the wild-caught bonobos living in zoos today came from wild populations that are most likely extinct."

Less accessible to many readers but equally important for Dr. Reinartz's conservation efforts is her academic work. Part of her doctoral thesis on the extent of genetic diversity in bonobos compared to that in chimpanzees was published in 2000 by the journal *Molecular Ecology.* In 2006, she joined Congolese researchers Inogwabini Bila Isia, Mafuta Ngamankosi and Lisalama Wema Wema in publishing a detailed examination of the effects of forest type and human presence on bonobo density in Salonga for the *International Journal of Primatology.* These and other publications also are a part of telling the world that there are bonobos in Salonga.

Sharing the Science

According to Frans de Waal in a 2005 interview when he was in Milwaukee to promote his 2005 book *Our Inner Ape,* both scholarly and popular publications and presentations are essential to protecting endangered species. Academic research is essential to keep and build the respect of the larger scientific community, he says, just as popular presentations are necessary to focus public attention and win widespread support.

Photo by Richard Brodzeller

Dr. Richard Carroll

Dr. Richard Carroll, director of the World Wildlife Fund's Office for Central Africa and Madagascar Programs, including the Congo Basin Forest Partnership, has hailed the Zoological Society's efforts as pioneering and lauded the solid science being done by Dr. Reinartz and her colleagues. That is one of the reasons the WWF, in partnership with UNESCO, decided to contract with the Zoological Society to support bonobo conservation efforts in the Salonga. Dr. Carroll said, "I wanted to support someone who was doing active work in bonobo land. Gay was getting things done on a very practical level. I really saw that she was the leader, both in captive bonobo conservation and in field work to protect the bonobo in the wild." Dr. Carroll invited Dr. Reinartz to a gathering of 160 experts to develop a cohesive conservation plan for the Congo River Basin. From that gathering in April 2000, the Congo Basin Forest Partnership was formed in 2002. As of 2006, the partnership had grown to 10 African countries, which, he said, "have seen that conservation is the way of the future and a way to get international attention."

Cartoons for Conservation

From the beginning, Dr. Reinartz has pursued a two-pronged approach to the Society's work, using both scholarly and popular avenues to build support for bonobo conservation. One of the most effective tools in Congo has been a

Bonobos: Encounters in Empathy

Comic-book-style magazines explain the natural history of the bonobo.

relatively modest color cartoon brochure produced by Bleu Blanc, a Congolese non-profit education group printing magazines for schoolchildren. Created by Delfi Messinger, the booklet tells in Lingala (lin-GAH-lah), one of the DRC's four national languages, the story of a tribal elder who surprises his hosts by refusing a dish of bonobo meat. The elder explains that he refrains from eating bonobos because, in his youth, the primates had saved him and his mother from ambush by an enemy tribe. During the war, 40,000 copies of the brochure were printed with funds from the U.S. Agency for International Development and were distributed to areas having military outposts where soldiers were taking baby bonobos as pets and were engaged in the shooting and selling of bushmeat.

As she said in her joint presentation with Inogwabini, "Our aim is to build the capacity of the Congolese to become involved in bonobo conservation and monitor bonobo populations as well as protect them."

At the research station at Etate, that work is well under way.

A Day at Etate

A foray into the forest by the Zoological Society's research team.

CHAPTER 8.

A Day at Etate

Long before first light filters through the rain-forest canopy, the work day at Etate Research Station begins with song. Here on the west bank of the Salonga River in the Democratic Republic of Congo, humans begin to stir at about 4:30 a.m. On the other side of the world, back home in Wisconsin, it is still yesterday. In her small tent, Dr. Gay E. Reinartz can hear soft singing as the men begin their day with hymns and chanted prayers. The words are foreign. The men sing and pray in their native tongues or the trade language Lingala. But often the melody is familiar. "How Great Thou Art" – sometimes extended with verse after improvised verse – is a favorite.

One of the Etate guards takes up a bundle of palm fronds and, with a rhythmic swish, begins to sweep patterns in the sandy yard of the encampment. A few late sleepers snore. Children talk and giggle. A hungry infant cries. Relatives occasionally visit the research station. On some occasions, even Dr. Reinartz has a family member in camp. Several times, she has been accompanied by her son, Nathaniel, who supports the conservation missions with hard work, long days and a willingness to pay for the privilege. Someone jumps into the river for a quick wash. A clatter of pots announces that Loketshi Lokuli Mira, the camp boss and chief of logistics, is preparing breakfast of coffee and fried plantains.

photo provided by Gay E. Reinartz.

Dr. Reinartz uses a solar panel to charge up equipment for a day at the Zoological Society's rain-forest research station, where there is no electricity.

After dressing, Dr. Reinartz switches on her headlamp, illuminating the thatched hut that serves as a dormitory and shelters the tents. Sitting cross-legged outside her tent, she begins to organize the day. She is surrounded by her office: clipboard, satellite phone, power pack to recharge the phone and the solar panel that renews the power pack, account books, stacks of data sheets on each day's observations, lists of things to take into the forest.

"NOBODY GETS UP ACTING GRUMPY"

Reinartz emerges from her tent with a pleasant smile and a courteous "Bonjour!" So does everyone else.

"Nobody gets up acting grumpy," she says. "It is simply not done."

No matter how tired, no matter how ill (malaria is always a threat, and Dr. Reinartz suffered a bad bout on her first trip to the Salonga), the residents of this small enclave greet every morning with courtesy.

People sit around the campfire as they eat. There is time for what Dr. Reinartz calls kind conversation. Although her role is to drive the day's activities, she is careful not to rush. A soft approach is valued. As night fades to gray about

5:30 a.m., she asks of those wise in the ways of rain-forest weather, "Will it rain?"

She inquires of Mira about his plans. If this is a day they will have meat, he wil] kill and pluck the chicken. Bobo, Mira's assistant, will do laundry. Once th(research party arrives at the station, there is little occupation for the two arme(soldiers whose presence has been necessary on the remote river stretches als(traveled by poachers and rogue military bands. So the guards join in the daily wor] of keeping the research station going – cleaning latrines, gathering firewood refilling water bottles from the river and disinfecting the water supply with iodin(A less frequent and much prized assignment, viewed as a virtual holiday, is t(take a boat to locate, bargain for and bring back food from the nearest village, a] excursion of five to six hours.

Dr. Reinartz discusses the forest area they plan to survey with Mboyo Bolinga, chief of operations at Etate; her counterpart, Conservateur Botomfie (buh-TUM-fee) Mompansuon of l'Institut Congolais pour le Conservation de la Nature (ICCN); and with Professor Lubini Constantin (loo-BEE-nee CON-stan-teen), professor of botany at the University of Kinshasa. Prof. Lubini, an expert in the forests of equatorial Africa, joined the research team for its spring mission in 2004.

Photo provided by Gay E. Reinartz.

Prof. Lubini Constantin of the University of Kinshasa identifies leaves for the Zoological Society.

Duties in the camp settled, Reinartz and her team load up with machetes, water, field note-books, cameras and sample bags, and set off into the surrounding jungle.

"I like actually to be out in the forest by 6 o'clock," she says. "We have a bette] chance to see bonobos."

Just getting to the start of a new survey trail often is a journey of several miles b] foot or pirogue. Today they will walk a little less than 2 miles in dim light on slip pery forest paths. Moisture is constant even when it doesn't rain, and there ar(only about 20 days a year without rainfall. Plants grow so quickly here tha overnight branches and vines have intruded on the path cleared only yesterday.

In addition to Dr. Reinartz, Conservateur Botomfie and Prof. Lubini, today's fiel(team includes her research assistant, Guy Tshimanga (shih-MAHN-gah), who is a graduate student at the University of Kinshasa. Guy will record most of the fiel(data and get experience using the Global Positioning System recording the longi tude and latitude of the team's position. The team is led by two local men whos(expertise as naturalists has emerged from years of working together with Dr. Reinartz in the rain forest, Mboyo Bolinga and Isomana Edmond.

INTO THE GREEN TWILIGHT

Sure-footed, forest-wise Edmond leads into the green twilight.

"If I can watch Edmond walk and I can step where he steps, I don't fall," say Dr. Reinartz.

As he nicks new growth from the trail with his machete, Edmond watches fo] snakes, signs of leopards, other possible dangers.

Bonobos: Encounters in Empath}

"If it's early enough, we hear monkeys calling," says Dr. Reinartz. "That slows us down because we stop and try to get a count of individuals."

Photo provided by Gay E. Reinartz.

Researchers identify a tree limb bent by a bonobo.

On occasion they hear bonobos, although seeing them is rare. Sometimes they reach a nest moments after the residents have left.

"You can even smell them," she says, "but we don't see them often."

Once they reach higher ground, they traverse one or two of the transects cut by previous research visits, each nearly a mile long. On this mission, the goal is matching scientific with colloquial names of about 40 species of trees in which bonobos build nests. Collecting specimens for identification has proved chancy. Dr. Reinartz returned to Etate one spring to find that termites had eaten all the plant presses. She later learned that smoking the plant presses over the campfires kept out the insects.

Having Prof. Lubini present to identify the trees on the spot is invaluable, as is the forest savvy of Mboyo Bolinga and Edmond. Together, everyone learns. Prof. Lubini becomes acquainted with the vernacular names used by the Mongo people and gets to study trees limited to this region of the country. Mboyo Bolinga learns the scientific names of trees, a proud accomplishment. He carefully records each Latin name into his notebook. Among Salonga's 300 or 400 kinds of trees, differences can be subtle. Sometimes identification hinges on a single gland of a leaf, and the canopy can be as high as 100 feet overhead. The two local naturalists show the professor not just adult trees soaring out of sight, but also nearby saplings of the same species, with leaves within reach for close scrutiny. At first Prof. Lubini is reluctant to trust the identifications of Mboyo Bolinga, who is without university degrees, but as the mission progresses, identifying trees day after day, the professor more often consults the younger man.

Work continues with stops only for water, not for food, until about 3 p.m.

"We go as far as we can go in one day," Dr. Reinartz says. "We do as much as we can do. I take stock. We're all spent. We hightail it back to camp to arrive before it gets dark."

Speeding Past Plastic Streamers

Encouraging themselves by counting off the yellow or pink plastic streamers of forestry tape that mark the distance of their trails, the researchers speed along through air thick with humidity. Night falls quickly in the forest, usually by 5:30 p.m., and darkness makes it difficult to follow the machete-cut trail. So research teams try to beat nightfall back to camp.

"It's hot and steamy," Dr. Reinartz says, "unless it rains. Then it's cold, cold, cold."

As they near camp, they smell dinner – mountains of rice, sometimes enhanced by chicken or fish from the river. Vegetables are infrequent additions. They are hard to get locally.

Dr. Reinartz says, "The work is physically very demanding, and we have such a short amount of time at Etate that I have this idea that we have to make use of every second. We go and we just kill ourselves, but we can rest when it's over. The only way I can drive this hard is to set the example. I'm usually the oldest person hiking in the woods; so the leverage is equal to my age."

When they reach camp, there will be coffee or tea waiting. Then Dr. Reinartz retreats to her thatch hut bathhouse and slops herself with cold water from a bucket, the Etate version of a shower. Everyone reassembles for dinner. Some linger around the fire. As they talk, the men take turns using machetes with scalpel-like precision to cut open lumps on each other's feet. The lumps, having black centers, contain parasites, small larvae with black heads that burrow into the skin. Dr. Reinartz will wait until she returns to her tent to perform a similar operation on herself with a pocket knife. She usually stays for part of the campfire conversation. She is working to build a team as well as a research station and some have been less open than others to her efforts. She makes it a point to include the pirogue pilots and the military guards in the daily field excursions so they, too, have a chance of seeing (or at least hearing) for the first time the elusive primates at the center of the all the research and conservation work.

"It just took a lot of time for Moyembi (moh-YEM-bee) Vincent to accept me," she says of one of the river pilots from Mbandaka. "It wasn't until he had a cut on his ankle that got infected and I cleaned it out and patched it that he started to thaw. But bit by bit, they see you for who you are. They see you're just here doing your job. You're not overworking them. You're not cheating. A deal is a deal. Then before you know it, you've got a team."

"WE ARE ALL BEHIND YOU 100%"

Loyalty to the common goal has grown with each mission, to the point that the skeptical Vincent, now a full-time pirogue pilot for the Zoological Society, told her on a recent trip: "Madame, you know we are all behind you 100%."

On the way back to Kinshasa on this particular mission, Prof. Lubini observes, "I'm going to miss it. We've had a family. For four weeks we've had a family."

The daily data sheets await her entries. Dr. Reinartz retreats to her tent, turns on her headlamp and begins filling out the daily log. She plans the next day's mission and then falls asleep to the murmur of deep voices.

Photo provided by Gay E. Reinartz.

Dr. Reinartz (in back) and the Zoological Society's research team (from left): Nduzo Bokono-Bolungi, Mboyo Bolinga, Botomfie Mompansuon, Isomana Edmond.

Bonobos: Encounters in Empathy

Partners in Preservation

Survival of wild bonobos and their relatives in captivity are interlinked. Laura and infant Claudine (foreground photo by Richard Brodzeller) are superimposed over a photo of a bonobo in Congo (provided by Gay E. Reinartz).

CHAPTER 9.

Partners in Preservation

Gay E. Reinartz has a quick, consistent reply when people suggest that bonobos need "another Jane Goodall to promote the species." She responds in the much same way she did in May 2000, when she and Inogwabini Bila Isia, then field director for the Zoological Society's conservation program in Congo, addressed the Great Apes Conference on Challenges for the 21st Century:

"What the bonobo needs is for us to help create the conditions (economic and social) in the Congo so that the next 'Jane Goodall' is black – a Congolese celebrity for conservation."

While that person is emerging, there is plenty for others to do.

The Democratic Republic of Congo, with an area about one-quarter of the size of the United States, is rich in natural resources – timber, petroleum, diamonds, gold and coltan, a mineral used in cell phones. Like all countries so endowed, the nation is subject to the disadvantages of economic wealth based on extraction and eventual depletion of those resources. The DRC is party to international treaties on biodiversity; climate change; desertification; endangered species; hazardous wastes; law of the sea; marine dumping; and protection of the ozone layer, tropical timber and wetlands. In the spirit of a proud history of conservation, Congolese government officials recently pledged to protect 15% of the country's land, almost doubling the 8% already protected by law.

Salonga National Park, which was created in 1970 specifically to protect bonobos, is the second largest tropical forest park in the world, covering a gigantic 13,896 square miles of rain-forest and savanna forest mosaic. That is almost the size of Massachusetts and Connecticut together and more than four times bigger than Yellowstone, the largest national park in the United States.

Dr. Reinartz notes that timber interests hold logging concessions or harvest permits in 40% of the bonobos' traditional range, and the logging industry is viewed as a potential source of much desired economic development. This is a nation where annual income averages something less than $120 per person. Life expectancy of 50 years for men and 53 for women is shorter than that of the oldest captive bonobos. The DRC's population of 62.6 million is growing at a rate of 3% a year.

Photo provided by Gay E. Reinartz.

The fork in two rivers (Salonga on left, Yenge on right) marks the entrance to Salonga National Park. There is no sign.

HUNTERS EVERYWHERE

As Dr. Reinartz points out, most of the indigenous people look to the forest not in hopes of seeing bonobos, but to find food, shelter and fuel for themselves and their families. Just as hunting for the bushmeat trade poses the most acute danger to bonobo survival, commercial logging and deforestation for agriculture form chronic threats as habitat disappears. Writing with three Congolese co-authors for the *International Journal of Primatology* in 2006, Dr. Reinartz reported that in their study they found no evidence of bonobos in some areas of the Salonga National Park where suitable habitat existed. Even more ominously, "no sample location was free of evidence of hunting." However essential conservation may be to our long-term survival, in the short run, conservation is a luxury good, something that most people will support only when their own immediate survival is assured. Therefore, richer nations must provide sufficient financial incentives to make conservation something other than a sacrifice.

Not even the massive size and remote location of Salonga National Park can ensure protection for the bonobo, the forest elephant or any of the other species living there, Dr. Reinartz and her colleagues say. Although the park is larger than the country of Rwanda, it is by no means remote enough to limit human incursions. The park, not all of which is forested, is divided by six major river systems, all of which provide access. Many forest trails through Salonga are used as major footpaths for people who live near the park's boundaries. Although the Zoological Society has staffed three anti-poaching patrols based in Salonga, there are not enough patrols along most rivers and pathways to control illegal access. Therefore these routes serve as doorways to further exploitation of forest plants and animals, bonobos included.

ALWAYS MORE ORPHANS

There is no more poignant proof of this ongoing danger than the continued appearance in Kinshasa of orphaned bonobos whose parents have been shot to provide meat and to allow capture of their infants for sale. No matter what laws exist to restrict this trade, humans persist in keeping "the cute little monkeys" as pets. A badly dehydrated young female was confiscated Dec. 23, 2005, at Charles de Gaulle Airport near Paris in the hand luggage of a Russian traveler from Kinshasa. Fortunately, the infant – who was rubbed raw by a chain around her waist – survived the return trip to Kinshasa. There she joined 46 other orphans being cared for at *Lola ya Bonobo* (Paradise for Bonobos), a sanctuary near the DRC capital. The rehabilitation center, estab-

A bonobo orphaned when its mom was killed by poachers sits in a Mbandaka doorway.

lished in 1994 by director Claudine Andre, is a charter member of the Pan African Sanctuary Alliance and is operated by Friends of Bonobos in Congo (*Les Amis des Bonobos du Congo*) in collaboration with the nation's Ministry of the Environment.

Although work at Etate, in one sense, is just beginning, Dr. Reinartz and her co-authors have concluded that Salonga National Park indeed is home to a substantial bonobo population whose distribution is patchy throughout the land scape. Their research has progressed far enough to support other important observations as well. Their data suggest that the type of forest, its size and the consistency of forest type all affect how many bonobos live in a given area of the park. Adult bonobos build a nest in the canopy each night. The density of bonobo nests corresponds to the presence of mature forests of mixed species that have an understory of herbaceous plants belonging to the Marantaceae family. Nest groups are larger where the forest provides abundant supplies of food and preferred nesting trees. Therefore, the patchy bonobo distribution corresponds with variations found in the habitat and with areas where frequent illegal hunting occurs.

Dr. Reinartz and her co-authors make it clear that their estimates are only preliminary and their quest for better information continues; yet they acknowledge the utility of a "best-guess estimate." Further protection for the park is needed now, and they already have enough evidence to advocate a policy of stronger safeguards. Using nest density and proportion of area covered by nesting habitat as a basis for their calculations, the researchers project that Salonga could host a maximum of 19,000 adult bonobos. It is likely that there are far fewer than that number currently living in Salonga, due to poaching, researchers warn. To understand the dynamics affecting that population and to preserve the bonobos, conservation funding is necessary.

INVISIBLE TREASURES

Milwaukee's 21 bonobos also have immediate needs. Although their indoor exhibit area was especially designed to accommodate their preferred lifestyle, the outdoor enclosure is relatively small and almost completely inaccessible to Zoo visitors.

"Here we have this rich treasure, the zoo world's largest community of breeding bonobos," says keeper Barbara Bell, "and people can't see them when they are outside."

Milwaukee County Zoo Director Charles Wikenhauser wants to change that. He is pursuing county funds to begin planning the next phase of the Zoo's ongoing work to improve its facilities. Although pachyderms are high on the list of animals needing roomier homes, the needs of Milwaukee's stellar bonobo community also are being carefully considered in the Zoo's view of its future.

Wikenhauser also envisions a non-intrusive means of sharing with the public the interaction between keepers and bonobos that goes on during training sessions. However eloquent and well-informed, the volunteers who tell visitors what is happening when the bonobos are off exhibit can only explain. They can't show people what is happening. Right now, there is no chance for zoogoers to see for themselves how well keepers and bonobos work together. With the bonobos' vulnerability to human diseases, it isn't possible to allow the public to pay regular backstage visits, Wikenhauser says. However, he has an idea to work around that difficulty.

Photo by Richard Brodzeller

Barbara Bell examines baby Makanza's belly to get him used to being touched for veterinary exams.

He says, "If we had this on video and people could watch the training sessions on plasma screens like those we have in the feline building, people could see for themselves what we are doing."

Being host to the zoo world's largest captive community of an endangered species with as much potential star power as the bonobo is the kind of achievement that exhibits a zoo's true excellence, Wikenhauser says.

"It's great to have a species where you're the best," he says. "We get calls from all over who want to know, 'How do you do that?' Our keepers are really on the cutting edge of what's going on."

Milwaukee's experience with bonobos is proof of what success is possible when people who care pay attention to those in their care, Wikenhauser says.

Learning from the Animals

"We have been really good at learning from the animals," he says. "There may not be language, but there certainly has been communication. These are such wonderful animals, and it's an amazing story in itself about how many different kinds of people care about them. There's the animal care staff, of course, but also the docents and donors. Once they learn about (bonobo) social structure, people are just fascinated. We've gotten so much help from the medical community, too."

What pleases Wikenhauser most about the bonobo community in Milwaukee is its diversity. Too often, he says, institutions with captive-breeding programs have an approach that insists on having only the healthiest animals and on breeding them as often as possible. He feels that Milwaukee shows its strength in its willingness to take on problem animals – the old, those with chronic complaints such as diabetes, and the socially or psychologically maladjusted – in addition to healthy founder animals that have created a thriving colony.

Photo by Richard Brodzeller

Based on the success of the bonobo operant-conditioning program, elephants at the Milwaukee County Zoo were trained to present their feet for nail care.

"We've got the space and we don't mind tackling problems," he says. "In this way Milwaukee has been a benefit to the bonobos in captivity."

Bonobos, in turn, here have been a benefit to the 369 other species resident at the Zoo's 209 leafy acres on West Blue Mound Road on the western edge of Milwaukee. The success their keepers experienced in using training to transform what used to be a hostile relationship between humans and bonobos into a cooperative partnership was noticed immediately by keepers in other areas. Those not already engaged in training promptly developed programs appropriate for their own charges. Voluntary operant conditioning began to yield results all across the Zoo, Wikenhauser says

Whenever Dr. Reinartz speaks about the Zoological Society's conservation efforts in Congo, she invariably credits the role played by the Zoo's bonobos. To keep animals in the Zoo without engaging in a larger conservation effort to support and preserve wild populations is nothing more than exploitation, she says. "It all comes back to these animals," she says. "They are the genesis for the whole operation."

Indeed, bonobos at the Zoo have initiated considerably more than the transformation of their own living conditions and the preservation of their wild cousins. They have become focal points of an international examination of the development of human empathy conducted by Australian biologist and best-selling author Jeremy Griffith. As Griffith prepares a documentary on "The Human Condition," participating scientists, philosophers and ethicists have engaged in cyberconversations on the topic. Their consideration ranges from the biological basis for compassion and the well-spring of human consciousness to the evolutionary advantages of kindness. Milwaukee's bonobos are often at the center of the discussion.

Dr. Harry Prosen, who became involved in the project after Griffith became aware of his and Bell's observations at the Zoo, says, "In *The Descent of Man* Charles Darwin said humans' moral sense affords the best and highest distinction between man and the 'lower animals.' However, biologists have struggled to explain the origins of such an altruistic moral sense in our own make-up. In our studies of bonobos, could we be witnessing the formative stage of such a moral sense? It's an extraordinarily exciting prospect to consider that we may have found a window into the origins of our species' core nature."

The possibility that a close study of bonobos may be able to shed light on human origins indicates just how important an opportunity is being offered at the Zoo, he says. His observations with Bell at the Zoo are considered to be essential, if often controversial, contributions to the ongoing dialogue.

Dr. Prosen says, "I am constantly astounded by the degree of fascination there is among people around the world at what Barbara and I have seen and noted here in Milwaukee."

With a laugh, he chides himself for being surprised.

"After all," he says. "I ought to understand after all these years. I come in (to the off-exhibit area) and here are keepers hosing down enclosures, removing dirt, and 10 minutes later they are discussing in depth the most complicated intellectual question you can imagine."

Bell shrugs and says, "That's just the Zoo."

That is what good zoos like Milwaukee County's have become. During the last 30 years, Wikenhauser says, zoos have made the dramatic shift from being users of wildlife to being conservers of wildlife. That philosophical change has affected everything from the way animals are housed and exhibited to the way they are provided with adequate nutrition, physical exercise, mental stimulation and medical care.

"When I began work as a keeper at a small zoo in Rock Island, Illinois," he recalls, "the prevailing attitude was this: If an animal died, it was sad, but people assumed that the zoo would get a replacement. Now strong international agreements restrict the sale and import of endangered species of all kinds. Instead of raiding the remaining wild populations, zoos are actively involved in trying to save and even restore them."

Deputy Zoo Director Bruce Beehler provides a few examples: Guam rail chicks hatched at the Milwaukee County Zoo are being transported to Guam to increase

Photo by Richard Brodzeller

The Zoo, Zoological Society and Wisconsin Department of Natural Resources worked together to bring back trumpeter swans to Wisconsin.

the numbers of a bird species that teeters on the edge of extinction. Closer to home, trumpeter swans again are nesting throughout Wisconsin due to a restoration program in which the Zoo and the Zoological Society played a large part. Zoo researchers funded by the Zoological Society are continuing a long-term study of a breeding colony of Humboldt penguins off the coast of Chile. Zoo scientists are investigating a deadly fungus that is wiping out frog populations in Panama. The frogs are being treated with drugs purchased with funds raised by Zoo Pride volunteers. Blood samples collected regularly from the Zoo's black rhinoceros are being used in several medical and reproductive studies of this highly endangered species. Golden lion tamarins bred at the Zoo have been returned to the Brazilian rain forest to buttress the dwindling population of this increasingly rare monkey.

Photo by Richard Brodzeller

Blood samples from the Zoo's black rhinos are used in medical and reproductive studies to help this endangered species.

Dr. Beehler says: "Zoos focus our attention and direct our help toward bonobos and frogs and penguins and sharks and rhinoceroses and bats and tigers and many other creatures, great and small. Zoos bring together caring people – the visitors, the staff, the support organizations, the volunteers, the donors, scientists, other zoos, conservation groups, government agencies – who want to nourish and protect our planet's thin web of life."

Wikenhauser points to the strong partnership that has been formed by the Zoo, the Zoological Society and its partner, the Foundation for Wildlife Conservation, Inc.: "Together we have invested $75 million in capital improvements to the Zoo and thousands of dollars in conservation programs throughout the world. Together – through education programs conducted by the Society with content provided by the Zoo – we are spreading the word about what we as human beings can do to ensure that the diversity that makes our world so beautiful can continue to amaze future generations."

To Dr. Gil Boese, president of the Foundation for Wildlife Conservation, Inc., who has been a part of the story since the first seven bonobos were brought to the Zoo, conservation programs are the natural outgrowth of the concern that responsible zoos devote to the animals in their care. Similarly, he says, the partnership that gave birth to the Bonobo and Congo Biodiversity Initiative was bound to lead to other partnerships.

He says: "Like any of these programs, you start out by trying to conserve a species and you end up involved in the collective ecology of the region: the animals, the plants, the people.

"True conservation, as I see it, is not imposing conservation on the local people but getting them to become part of the conservation process. As they get more and more understanding of their own ecosystem, they also are incorporated into the ecosystem. In other words, they prosper in quality of life not at the expense of the

The Zoological Society delivered farming tools to Congolese to help revitalize agriculture in villages surrounding Salonga National Park, so that people will depend less on bushmeat.

environment but in becoming a partner with the environment. Obviously, funds have to come in to keep this going. We're giving people funding to DO things, to set up protections and research and education. We're assuming that by getting them to understand the value of bonobos to them, we'll reinforce the message not to kill the animals. With minimal impact, we get maximum benefits to both the environment and the people."

In a world in which most human beings will never see any of these animals alive in their natural homes, zoos play an increasingly important role in providing what Dr. Prosen would call opportunities for empathy. Human families visiting the Stearns Family Apes of Africa building can see on the other side of the glass bonobo families that are not only unmistakably different but also recognizably familiar.

Dr. Robert M. (Bert) Davis, chief executive officer of the Zoological Society, is proud of the fact that the organization he now heads has become an advocate for the least known of the great apes both in captivity and in the wild.

He says: "Our goal is to share the message: Save the bonobos. Their survival is important to our survival as human beings. Their make-love-not-war culture, their reverence for the aged, and their female-dominated society are changing how scientists think about human development. We think this (story) will show you that individuals can and do make a difference in the world. We hope it will inspire you to be one of them."

Human families visiting the Zoo encounter bonobo families that are recognizably familiar. In the foreground are Maringa (left) and Zomi.

Bonobos: Encounters in Empathy

For More Information *

BOOKS

Beck, Benjamin B. et al. *Great Apes & Humans: The Ethics of Coexistence.* Smithsonian Institution Scholarly Press, 2001.

Caldecott, Julian, and Miles, Lera. *World Atlas of Great Apes and their Conservation.* University of California Press, 2005.

de Waal, Frans. *The Ape and the Sushi Master: Cultural Reflections of a Primatologist.* Basic Books, 2001.

de Waal, Frans. *Bonobo: The Forgotten Ape.* University of California Press, 1997.

de Waal, Frans. *Good Natured: The Origins of Right and Wrong in Humans and Other Animals.* Harvard University Press, 1996.

de Waal, Frans. *Our Inner Ape.* Riverhead, 2005.

de Waal, Frans. *Peacemaking among Primates.* Harvard University Press, 1990.

Elwood, Ann, and Wexo, John Bonnett. *Chimpanzees and Bonobos.* Creative Education Inc., 1991.

Kano, Takayoshi, and Vineberg, Evelyn Ono. *The Last Ape: Pygmy Chimpanzee Behavior and Ecology.* Stanford University Press, 1992.

Marchant, Linda. *Behavioural Diversity in Chimpanzees and Bonobos.* Cambridge University Press, 2002.

Smith, Neil. *Language, Bananas and Bonobos: Linguistic Problems, Puzzles and Polemics.* Blackwell Publishers Ltd., 2002.

Susman, Randall L. (Ed.). *The Pygmy Chimpanzee Evolutionary Biology and Behavior.* Springer, 1984.

Tuttle, Russell H. *Apes of the World: their social behavior, communication, mentality and ecology.* Noyes Publications / William Andrew Publishing LLC, Park Ridge, N.Y., 1986.

ORGANIZATIONS MONITORING CAPTIVE BONOBOS

NORTH AMERICA

Association of Zoos and Aquariums (AZA), Species Survival Plans (SSP)
http://www.aza.org/CANDS/index.cfm?page=SSP_detail
Bonobo SSP coordinator: Dr. Gay E. Reinartz, the Zoological Society of Milwaukee: www.zoosociety.org, (select conservation/bonobo conservation)

EUROPE

European Association of Zoos and Aquaria (EAZA), European Endangered Species Programme (EEP)
http://www.eaza.net/EEP/3greatapes.html

Institutions Housing Bonobos

North America

Cincinnati Zoo & Botanical Gardens
Cincinnati, OH
www.cincyzoo.org (or www.cincinnatizoo.org)

Columbus Zoo
Columbus, OH
www.colszoo.org (or www.columbuszoo.org)

Fort Worth Zoo
Fort Worth, TX
www.fortworthzoo.com

Great Ape Trust of Iowa
Des Moines, IA
www.greatapetrust.org

Jacksonville Zoo and Gardens
Jacksonville, FA
www.jaxzoo.org (or www.jacksonvillezoo.org)

Memphis Zoo
Memphis, TN
www.memphiszoo.org

Milwaukee County Zoo
Milwaukee, WI
www.milwaukeezoo.org
Zoological Society of Milwaukee
www.zoosociety.org, (select conservation/bonobo conservation)

San Diego Zoo and San Diego Wild Animal Park
San Diego, CA
www.sandiegozoo.org

Zoologico Benito Juarez
Morelia, Mexico

Europe

(*Some sites offer English translations. For translated versions, search the zoos
in Google and click on "Translate this page" to the right of the Web site headline.*)

Apeldoorn Zoo
Apeldoorn, the Netherlands
www.abenteuer-zoo.de

Berlin Zoo
Berlin, Germany
www.zoo-berlin.de

Frankfurt Zoo
Frankfurt, Germany
www.zoo-frankfurt.de/index_e.htm

Koln Zoo
Cologne, Germany
www.zoo-koeln.de

Leipzig Zoo
Leipzig, Germany
www.zoo-leipzig.de
Lisbon Zoo
Lisbon, Portugal
Planckendael Zoo
Royal Zoological Society of Antwerp
Hofstade, Belgium
www.planckendael.be
TWYCross Zoo
Atherstone, Warwickshire, Great Britain
www.twycrosszoo.com
The Zoo Wilhelma
Stuttgart, Germany
www.wilhelma.de
Wuppertal Zoo
Wuppertal, Germany
www.zoo-wuppertal.de

CONSERVATION OF BONOBOS IN THE WILD
African Wildlife Foundation (AWF)
www.awf.org
Ape Alliance
www.4apes.com
Bonobo Conservation Initiative
www.bonobo.org
Bushmeat Crisis Task Force
www.bushmeat.org
CARPE (Central African Regional Program for the Environment)
http://carpe.umd.edu
Congo Basin Forest Partnership
http://www.cbfp.org/en/index.htm
Great Ape Project
www.greatapeproject.org
Great Apes Survival Project (GRASP)
www.unep.org/grasp/
Lola ya Bonobo the A.A.C. Bonobo Sanctuary of the Democratic Republic of Congo-Kinshasa
http://bonoboducongo.free.fr/us/
Lukuru Wildlife Research Project
http://members.aol.com/jat434/index
Max Planck Institute
http://www.eva.mpg.de/primat/files/bonobo.htm
United Nations Education, Scientific and Cultural Organization (UNESCO)
www.unesco.org
United States Agency for International Development (USAID)
Wildlife Conservation Society
www.wcs.org

World Wildlife Fund (WWF)
www.worldwildlife.org
Zoological Society of Milwaukee (ZSM) **and the Foundation for Wildlife Conservation, Inc.** (FWC)
www.zoosociety.org

DEMOCRATIC REPUBLIC OF CONGO INFORMATION
CIA World Factbook
www.cia.gov/cia/publications/factbook/

GENERAL BONOBO INFORMATION
ARKive, images of life on Earth
http://www.arkive.org/species/GES/mammals/Pan_paniscus/more_moving_images.html
BBC News*
http://news.bbc.co.uk/
Doug Foster, Focused on Discoveries in Animal Behavior Research as part of 2000 fellowship for Alicia Patterson Foundation
www.aliciapatterson.org/APF2002/Foster/Foster.html
Living Links Center, Emory University
www.emory.edu/LIVING_LINKS/
Milwaukee Journal Sentinel*
www.jsonline.com
Wikipedia*
www.wikipedia.org/
Primate Info Net of the Wisconsin Primate Center
www.primate.wisc.edu/pin/

MILWAUKEE COUNTY ZOO AND ZOOLOGICAL SOCIETY OF MILWAUKEE BONOBO PUBLICATIONS
(primary and/or co-authorship, or studies conducted with the Milwaukee County Zoo)

Bell, B., Clyde, V., Khan, P., and Maurer, J. 2000. Advanced Operant Conditioning and Reproductive Applications in the Bonobo *Pan paniscus*. The Apes: Challenges for the 21st Century. 373.

Clyde, V. 2006. Morbidity Review of Respiratory Infections in North American Captive Bonobos 2000-2005. Bonobo SSP Veterinary Advisor Report

Clyde, V., Bell, B., Khan, P., Rafert, J., and Wallace, R. 2002. Improvement in the Health and Well-Being of a Bonobo *Pan paniscus* Troop Through A Dynamic Operant Conditioning Program. 2002 Proceedings American Association of Zoo Veterinarians. 1:45.

Clyde, V., Bell, B., Wallace, R., and Roth, L. 2000. Cardiac Evaluation in Non-anesthetized Bonobos *Pan paniscus*. The Apes: Challenges for the 21st Century. 125-127.

Clyde, V., Roth, L., Bell, B., Wallace, R., Slosky, D., and Dolan, J. 2002. Cardiac Gestational Ultrasound Parameters in Non-Anesthetized Bonobos *Pan paniscus*. 2002 Proceedings American Association of Zoo Veterinarians. 365-368.

Bonobos: Encounters in Empathy

Hutchins, M., Smith, B., Fulk, R., Perkins, L., Reinartz, G., and Wharton, D., 2001. Rights or Welfare: A Response to the Great Ape Project. In B. Beck (ed.), Great Apes and Humans: The Ethics of Coexistance. The Smithsonian Institution. Pp. 329-366.

Janssen, D., Clyde, V., Lowenstine, L., Killmar, K., Morris, P., Rideout, B., Oosterhuis, J., Sutherland-Smith, M., and Lamberski, N. 2006. Medical Management of Severe Respiratory Disease in Bonobos *Pan paniscus* – Workshop Report. American Association of Zoo Veterinarians. 1-3.

Jurke, M.H., Sommovilla, R.H., Harvey, N.C., and Wrangham, R.W. 2000. Behavior and Hormonal Correlates in Bonobos. The Apes: Challenges for the 21st Century. 105-106.

Prosen, H., and Bell, B. 2000. A Psychiatrist Consulting At the Zoo: The Therapy of Brian Bonobo. The Apes: Challenges for the 21st Century. 161-164.

Reinartz, G.E. 2003. Bonobos. In: The Great Ape Project Census: Recognition for the Uncounted. GAP books, Portland, OR. Pp. 35-44.

Reinartz, G.E. 2003. Conserving *Pan paniscus* in the Salonga National Park, Democratic Republic of Congo. Pan Africa News 102: 23-25.

Reinartz, G.E. 1992. The Bonobo SSP: Beyond captive management. Bulletin of the Chicago Academy of Sciences 15 1:36.

Reinartz, G.E. 1984. Local breeding programs: Their effects on the genetic structure of the population. Regional Proceedings of AAZPA. Wheeling, WV. Pp. 61-67.

Reinartz, G.E., and Bila Isia, I. 2001. Bonobo Survival and a Wartime Conservation Mandate. In: The Apes: Challenges for the 21st Century Conference Proceedings. Pp. 52-56.

Reinartz, G.E., Bila Isia, I., Ngomankosi, M., and Wema, L.W. 2006. Effects of Forest Type and Human Presence on Bonobo Density in the Salonga National Park. International Journal of Primatology, 27(2): 603-634.

Reinartz, G.E., Bila Isia, I., Ngomankosi, M., and Wema, L.W. 2004. Assessment of bonobo densities in the Salonga National Park: the effect of forest types and human impact: Abstract. American Journal of Primatology 62 supp 1: 41-42.

Reinartz, G.E., and Friedrichs, S. 2005. Highlights of bonobo conservation in the Democratic Republic of Congo. *Zoo View* Winter 2005. Volume XXXVII 4: 6-7.

Reinartz, G.E., and Friedrichs, S. 2003. Survey and Protection of Bonobos in the Salonga National Park, Democratic Republic of Congo. *Bonobo Banner* 1 (1): 9-11.

Reinartz, G.E., Friedrichs, S., Ellis, L., and Leus, K. 2005. Bonobo (*Pan paniscus*) Master Plan: Recommendations for the Global Captive Population. Zoological Society of Milwaukee, Milwaukee, Wisconsin.

Reinartz, G.E., Guislain, P., Bolinga, M., Isomana, E., and Bokomo, N. (in prep). Ecological factors influencing bonobo density and distribution in the Salonga National Park: applications for population assessment. In T. Furuichi and J. Thompson (eds.), Developments in Primatology: Progress and Prospects, Bonobos Revisited: ecology, behavior, genetics and conservation. Springer, N.Y., U.S.A.

Reinartz, G.E., Karron, D., Phillips, R.B., and Weber, J.L. 2000. Patterns of microsatellite polymorphism in the range-restricted bonobo *Pan paniscus*: considerations for interspecific comparison with chimpanzees *P. troglodytes*. Molecular Ecology 9: 315-328.

Reinartz, G.E., and McLaughlin, S. 2006. Conservation in the Congo: The Zoological Society of Milwaukee Travels to the Heart of the Congo to Save the Bonobo. AZA Connect, October: 8-10.

Reinartz, G.E., and Mills, J. 1996. A case history for supporting field conservation: The Zoological Society of Milwaukee. *International Zoo News* 43/5 270: 293-298.

Reinartz, G.E., Vervaecke, H., and Ingmanson, E. (in prep). *Pan paniscus*. In J. Kingdon, D. Happold, and T. Butynski (eds.), Mammals of Africa Vol. 1: Vegetation, Climate and Geology of Africa, Evolution of African Mammals, Primates, Elsevier, Oxford, UK.

Teare, A., Bell, B., Kuhlmann, R., and Geanon, G. 1996. Ultrasonographic Measurement of Fetal Growth in a Bonobo *Pan paniscus*. Journal of Zoo and Wildlife Medicine. 274: 477-481.

Thompson-Handler, N., Malenky, R., and Reinartz, G.E. 1995. Action plan for *Pan paniscus*: Report on free ranging populations and proposals for their preservation. Milwaukee, WI: Zoological Society of Milwaukee.

Wallace, R., Bell, B., Prosen, H., and Clyde, V. 1998. Behavioral and Medical Therapy for Self-Mutilation and Generalized Anxiety in a Bonobo *Pan paniscus*. 1998 Proceedings AAZV and AAWV Joint Conference. 393-395.

This list of resources was prepared by Zoological Society intern Megan Ivers. This is not meant to be a comprehensive listing of sources for bonobo information.

About the Partners

The Zoological Society of Milwaukee, which is nearly 100 years old, has headquarters at the Milwaukee County Zoo in Milwaukee, Wisconsin, in the Midwestern United States. Founded in 1910, the non-profit Zoological Society has always focused on the "big picture," realizing that love of animals, conservation, education and zoos are intrinsically connected. Its specific mission is "to take part in conserving wildlife and endangered species, to educate people about the importance of wildlife and the environment, and to support the Milwaukee County Zoo." Its conservation programs extend from Africa to Central and South America. Its education programs reach an estimated 200,000 people each year in Wisconsin, in Africa and in Belize. It supports the Milwaukee County Zoo with more than $5 million each year, plus, from 2001 through 2008, more than $15 million in capital campaign funds to improve the Zoo.

The Foundation for Wildlife Conservation, Inc., is a partner with the Zoological Society of Milwaukee in education and conservation. Started in 1992, the foundation manages wildlife sanctuaries in Wisconsin and in the Central American country of Belize. The foundation has headquarters at the Milwaukee County Zoo and sponsors, with the Society, an international bird conservation-research-education project called Birds Without Borders-*Aves Sin Fronteras*®. The foundation also works with international non-governmental organizations to protect habitats for endangered species.

The Milwaukee County Zoo was founded in 1892. As of 2007, it had a collection of 1,778 animals from 370 species. Its staff are involved in wildlife conservation and research programs both on Zoo grounds and in the field, ranging from the island of Grenada to the coasts of Chile. The Zoo's extensive training program for its animals teaches them to participate in their own health care, so that examinations and treatments are less traumatic and animals live longer. The Zoo is an accredited member of the Association of Zoos and Aquariums. The Zoo, which is managed by Milwaukee County, has completed eight of nine major building projects begun in 2001 as part of a $30 million public-private capital campaign with the Zoological Society of Milwaukee. The last project will be completed in 2008.

DESIGNED BY
ZOOLOGICAL SOCIETY ©REATIVE DEPT.
Printed on recycled paper 3232C07